THE
ASTEROID
THREAT

D1637026

THE
ASTEROID
THREAT

DEFENDING OUR PLANET FROM
DEADLY NEAR-EARTH OBJECTS

WILLIAM E. BURROWS

 Prometheus Books

59 John Glenn Drive
Amherst, New York 14228

Published 2014 by Prometheus Books

Cover image © Triff/Shutterstock.com
Cover design by Jacqueline Nasso Cooke

Inquiries should be addressed to
Prometheus Books
59 John Glenn Drive
Amherst, New York 14228
VOICE: 716–691–0133 • FAX: 716–691–0137
WWW.PROMETHEUSBOOKS.COM

18 17 16 15 14   5 4 3 2 1

Library of Congress Cataloging-in-Publication Data

Burrows, William E., 1937– author.
    The asteroid threat : defending our planet from deadly near-Earth objects / by William E. Burrows.
        pages cm
    Includes bibliographical references and index.
    ISBN 978-1-61614-913-0 (paperback) • ISBN 978-1-61614-914-7 (ebook)
    1. Asteroids—Collisions with Earth. 2. Comets—Collisions with Earth.
3. Natural disasters. I. Title.

QB651.B87 2014
551.3'97—dc23
                                                                2013051026

Printed in the United States of America

*To all my students, American, Russian,
and in other parts of the world, undergraduate and graduate,
with profound and lasting gratitude for all that you taught me.*

"Six hundred thousand people died, and the local damage was more than a trillion dollars. But the loss to art, to history, to science—to the whole human race, for the rest of time—was beyond all computation. It was as if a great war had been fought and lost in a single morning. . . . . After the initial shock, mankind reacted with a determination and unity that no earlier age could have shown. Such a disaster, it was realized, might not occur again for a thousand years but it might occur again tomorrow, and the next time, the consequences could be even worse . . . Very well; there would be no next time. *So began Project SPACEGUARD."*

—Arthur C. Clarke
Rendezvous with Rama

CONTENTS

PREFACE

Earth lives in an abidingly tough and dangerous neighborhood, but it is not doomed. Astronomers say that the next planet-buster—the one that could cause a life-threatening catastrophe—will not be back for a century. We will be ready for it.

The word *threat* was chosen for the book's title because that is precisely what it is. And it is a threat that can be mitigated (as the growing planetary-defense fraternity calls it) because, for the first time in the home planet's history, Earthlings have the wherewithal to stave off the means of their extinction because they have access to space. That allows for an aggressive defense in which a potential impactor can be either nudged off course two decades or more before it collides with Earth or attacked with the same kind of weapons that were designed for the Strategic Defense Initiative, also known as Star Wars in the 1980s. Another way to assure our survival is to spread out and homestead other worlds, starting with the kind of lunar and Martian colonies prophesized by this book's honorary godfather, Sir Arthur C. Clarke. Like that great visionary, I am an inveterate optimist who is unalterably convinced that Armageddon can be prevented and that survival and growth are both imperative and achievable.

WEB

1

CHELYABINSK: RUSSIAN ROULETTE

I t is ironic that Chelyabinsk takes its name from an eighteenth-century fortress, Chelyaba, and that it lies in an area where top-secret nuclear-weapons research was once conducted and where there was a serious nuclear accident at a reprocessing plant northwest of the city in 1957. The radiation that blanketed the region killed or permanently sickened many people, though not in Chelyabinsk itself. The city's one-million-plus inhabitants thanked God that they were spared. And they had reason to thank him again on February 15, 2013, when the meteorite struck.

Chelyabinsk is cradled in the low Urals, just to the south of Sverdlovsk and north of Kazakhstan, the latter having ancient ties to Russia's old nemesis, Turkey. Its remoteness made it the Kremlin's choice as one of the Soviet Union's main military production centers, and that included the "Mayak" atomic-weapons complex eighty kilometers north of the city, where nuclear weapons were long produced.

Accidents, nuclear-waste disposal, and the day-to-day operation of the large reactor and the radiochemical plant combined to poison the region. Nuclear-waste dumping in the Techa River in the 1950s caused so many deaths and so much disease that twenty-two villages along the river had to be evacuated. On September 29, 1957, a liquid-nuclear-waste storage-tank explosion at Mayak released plutonium and strontium into the environment, amounting to twice as much radiation as was released in the infamous accident at the Chernobyl Nuclear

Power Plant in Ukraine in 1986. The city's older inhabitants still vividly remember Mayak since it made their city of more than one million the most radioactively contaminated place in the world. The accident was, of course, ordered by Moscow to be kept secret.[1]

Chelyabinsk is also a major producer of military equipment and weapons, which, combined with the Mayak facility, made it a priority target for U-2 spy flights. Many of the area's old-timers also remember the May Day three years after the accident at the reprocessing planet, when an American U-2 reconnaissance plane flown by Francis Gary Powers was shot down over Sverdlovsk, two hundred kilometers due north. Powers and the exposed film from his aircraft were taken to Moscow and shown to the world by the wily Nikita Khrushchev to embarrass Eisenhower, the American president.

In common with other Russian cities of any consequence, and with metropolises around the world, Chelyabinsk has a top-ranked professional hockey team whose name reflects the city's self-image: *Трактор*, or *Traktor* ("Tractor" in English). In fact, half of the city's twenty-two most notable personalities were hockey players in February 2013, while two more were nationally known musicians (one of them also a noted composer), and one, Maksim Surayev, was a cosmonaut who logged six months on the International Space Station as a flight engineer and did a spacewalk with fellow cosmonaut Oleg Kotov. Surayev's feat did not win him the prestigious Hero of the Russian Federation medal (which has replaced the venerable Hero of the Soviet Union medal), perhaps because his flight was not on *Mir*, the accident-plagued Russian space station that was intentionally sent out of orbit and plunged to a fiery end into the ocean in 2001. In any case, the snub by the defense ministry caused a furor in *Roskosmos*, the nation's rightfully proud space agency. The accolade, which is one of Russia's highest honors, brings housing, pension and travel benefits, and a state funeral. The award has been granted to nearly every first-time cosmonaut since Yuri Gagarin, the first man in orbit, won it for his 1961 flight.[2]

Chelyabinsk has other distinctions that make it one of the country's major metropolises. It produces bottled water and a variety of sparkling and regular wines, and it boasts a Holiday Inn with all the amenities that are found in the chain's other hotels around the world, including air conditioning, a restaurant, a sauna, minibars, a gym, and Internet access. The city is proud of a three-line subway system, which has long been under construction, and Arbat Street, a wide, tree-lined avenue with benches, clothing and jewelry shops, bookstores, a photo gallery and even an Irish pub. Strolling up and down Arbat Street and other thoroughfares is encouraged by the city fathers, who studiously ignore the fact that Chelyabinsk is blanketed with smog so thick that it rivals Beijing as a danger for asthmatics and others with respiratory problems. And as a sure sign that at least some of its citizens are determined to shake off all vestiges of the Communist era and embrace America, they have even opened The Wall Street Café. Mindful of their sophistication and their city's status as a major urban center, they count their blessings and regularly attend church, which, with the exhilarated embrace of capitalistic profiteering, is another sign of the repudiation of Communism.

At the same time, a grim reality haunted Chelyabinsk. The harsh working environment, the accidents, the lethal radioactivity, and then the explosion at the waste-storage site all made the city socially as well as environmentally contaminated. Like other industrial cities the world over, its people's daily battle for survival in a tough and constantly challenging physical environment caused and perpetuated anger that often set neighbor against neighbor and kept the city in a state of low-level conflict with itself. Ballads and some fiction notwithstanding, there is nothing romantic about poverty, sometimes terrible working conditions, and pervasive illness that cannot be prevented or even treated for lack of adequate resources. And the citizens of Chelyabinsk were acutely aware of the degradation of their environment. It affected their dispositions, as it does in other places where people are impoverished, exhausted, and depressed by their surroundings. "The city is a place where people always seem bitter with each other," said music teacher Ilya Shibanov.[3]

There is one distinction, however, that Chelyabinsk emphatically does not savor. On February 15, 2013, the citizens of the city who believed in God suddenly had reason to think that he either was not watching over them or had launched a terrible weapon at them from heaven for a reason no one could fathom. At 9:20 that morning, people on Arbat Street and elsewhere heard a thunderous roar coming from the sky. When they looked up, they saw a large, bright, fiery object coming from the southeast and trailing a long, thick plume of smoke.

There was a blinding white flash so bright that the stark shadows of buildings slid swiftly—and sickeningly—across the ground. The massive explosion instantly turned yellow. Then it darkened to orange. The asteroid—which had become a meteor when it plunged into the atmosphere and then meteorites when the wall of air broke it into large fragments—was about fifty-six feet in diameter, weighed more than seven thousand tons, and was made of rock that was probably laced with nickel and iron. Since its velocity was a scorching forty thousand miles an hour, it took only a few seconds to penetrate most of the thickening air. It was the terrific speed that made the sonic boom that got the attention of the people below.[4]

The explosion sent out a blast of energy that was later estimated at 450 to 500 kilotons, which is almost thirty times the force of the atomic bombs that turned Hiroshima and Nagasaki into smoking, radioactive rubble. The shock wave that hit Chelyabinsk lasted for thirty-two seconds, which seemed interminable to those who suffered through it.[5] It set off car alarms, and mobile-phone networks became overloaded almost instantly as frantic people tried to find out what had happened and connect with friends and loved ones. It also sent dishes and glasses flying off the shelves in thousands of apartments, and it slammed into the Railway Institute (a school) with such force that it blew in windows. The thunderous noise sent students running to the windows to see what had happened. Like some of their elders, many thought that war had broken out and that the explosion was made by an American ballistic missile. Several of the youngsters were hit by flying glass, some cut seriously. Going to windows to see what had happened was understandable, but they were the worst places to be. Terrified and, in many cases, bleeding, the students made their way down a staircase that was thickly blanketed with glass and onto the street. There, hundreds of people stood, looking in awe at the sky. Office buildings were evacuated and classes in the schools were cancelled. At least twenty children were injured when the windows of another school and kindergarten shattered two minutes after the meteor exploded.[6]

It was the same in the Sverdlovsk, Tyumen, and Orenburg Oblasts; in the Republic of Bashkortostan; and in the northern reaches of Kazakhstan. In fact, to a degree that was imperceptible elsewhere because of Earth's ability to absorb punishing blows, the whole planet shuddered like a large animal that had been hit by a high-caliber bullet. The Comprehensive Nuclear-Test-Ban Treaty Organization reported that the sound wave registered on sensors from Greenland to Antarctica, making it the largest ever detected by its network.

"I opened the window from surprise," a woman named Darya Frenn later recalled in a blog. "There was such heat coming in, as if it were summer in the yard, and then I watched as the flash flew by and turned into a dot somewhere over the forest. And in several seconds there was an explosion of such force that the window flew in, along with its frame, the monitor fell, and everything that was on the desk." Frenn's mind began to race with grim possibilities. She, too, thought that another world war had started and that consequently there would be nuclear radiation, which she knew was deadly. Her hands and feet went numb as she waited to hear about the fate of her family. Then it suddenly became absolutely quiet again. Within an hour, Darya Frenn learned from her neighbors about the smoking thing that had exploded in the sky. The force of the explosion was deeply frightening to those who saw and felt it because it showed them, in the most dramatic way, how vulnerable living creatures and their habitats are to the indomitable violence of nature. "God forbid that you should ever have to experience anything like this," she said, still clearly shaken by what she had seen and heard.[7] Alyona V. Borchininova, a barmaid, was shaken, too. "It was eerie," she told an American journalist the next day. "So we stood there. And then somebody joked, 'Now the green men will crawl out and say hello.'"[8]

The flash from the explosion was brighter than the Sun, according to several people who saw it. "It was a light which never happens in life. It happens probably only at the end of the world," wrote Valentina Nikolayeva, a school teacher who was horrified by a sudden vision of what she thought was the apocalypse, the absolute end of everything. "A shock wave. It was not clear what it was, but we were deafened at that moment. The window glass flew." Since the outside temperature was five degrees Fahrenheit, she and everyone else with broken windows scrambled to cover them with anything they could find, including wooden boards and blankets. When emergency officials announced that a meteor had exploded, it occurred to Ms. Nikolayeva that it could happen again. "I am at home, whole and alive," she wrote. "I have gathered together my documents and clothes. And a carrier for the cats, just in case . . ."[9]

Denis Kuznetsov, a twenty-three-year-old historian who was in the outskirts of the city, said that he heard the explosion and felt the shock wave even where he was. First there was a blinding flash of light lasting several seconds that made him close his eyes. The light was "like ten suns," he said, adding, "This is no exaggeration." Then he said he experienced what felt like a "push" as a sound wave passed through his body. "For some seconds I simply stood" as he listened to the sound of breaking glass. After calming his parents, he tried to call his friends, but the cell phone was dead, so he went on the Internet and was able to reach a friend in the city center who told him that emergency responders were in the streets. At first there was widespread confusion, he said, with many people believing that a satellite had come down or that a plane had blown up. But within an hour, news broadcasts reported what had actually happened. "There was no panic," Kuznetsov said. "All behaved quietly."[10]

And some of the city's one million residents quietly noticed that weird things had happened. Glass jugs were said to have exploded, dishes cracked, and electronic equipment died behind unshattered apartment windows. Balconies rattled. And one man said that a bottle broke as he was holding it. Anna V. Popova was at home with her daughter when she saw the flash and then heard the explosion. She said that the destruction seemed to be illogical. The windows of her enclosed balcony were blown in, but her neighbor's identical windows had withstood the blast. "A lot of people suffered, not us alone," she said. "Who are we supposed to blame for all of this?" she asked rhetorically. "Nobody of course."[11]

Mikhail Yurevich, the governor of Chelyabinsk Oblast, immediately announced that keeping the city's central heating system operating was of paramount importance. He estimated that damage from the explosion came to more than one billion rubles ($33 million).[12]

By February 18—three days later—1,491 people, including 311 children, had requested medical assistance, with 112 hospitalized, two of them in serious condition. A fifty-two-year-old woman had to be flown to Moscow for treatment of a broken spine.[13] And by March 5, almost three weeks later, more than 7,200 damaged buildings were reported in Chelyabinsk and the surrounding area, including 6,040 apartment blocks, 293 medical facilities, 718 schools, 100 cultural organizations, and 43 sports facilities, one of them being the Traktor Sports Palace. Since the damage was extensive but, with few exceptions, not heavy, nearly all the structures were quickly repaired. The city promptly announced that all broken windows in apartment buildings would be replaced free of charge. That provided some consolation.[14]

Residents found further consolation after learning that the situation could have been much worse. For one thing, the meteor that blew up virtually over their heads was not the only large so-called near-Earth object (NEO) in the sky that day. But this second object posed no danger to them. About sixteen hours after the initial explosion, an asteroid with a diameter estimated to be ninety-eight feet and weighing forty thousand tons passed Earth at a distance of 17,200 miles, which is well inside the Moon's orbit.[15] It had been spotted a year earlier by astronomers at the Observatorio Astronómico de La Sagra in Granada, Spain, and was named 2012 DA14. They calculated that its next close approach to Earth would be on February 16, 2123, when it comes within nineteen thousand miles of the planet's core. At that distance, an impact is not expected.[16] But gravitational attraction being what it is, every pass will bring it closer, so a day of reckoning will probably come unless it is stopped before then. When it passed by in 2013, 2012 DA14 posed no danger to Earth, but it was a reminder that very large rocks are flying all over the neighborhood all the time; as the late Eugene M. Shoemaker, a geologist and the first de facto planetary scientist, said, NEOs in the vicinity of Earth amount to a "hail of bullets."[17]

Those in the Chelyabinsk region who knew Siberia's history had yet another reason to be consoled: that is, it was not Tunguska all over again. At a little after seven o'clock on the morning of June 30, 1908, the largest impact "event" in recorded history happened near the Podkamennaya Tunguska River, some 1,200 miles northeast of Chelyabinsk. The asteroid's size has long been debated, but it is estimated to have been roughly one hundred feet or less in diameter and to have exploded at an altitude of about five miles. The explosion seems to have been as powerful as a medium-size hydrogen bomb and at least several hundred times more powerful than the atomic bombs that were dropped on Hiroshima and Nagasaki. The airburst flattened tens of millions of trees over an area of about eight hundred square miles. A few eyewitnesses reported that trees snapped forty miles from the impact site and that a precedent was set when the shock wave smashed windows. There were no reported injuries.[18] Because of its magnitude, it attracted scientists for decades; more than one thousand scholarly papers were written about it, mostly in Russia. Calling such a destructive explosion an "event" is an understatement on the order of calling an attempted assassination an "incident."

"This is a much smaller event" than Tunguska, Tom Bissell, a journalist and a fiction writer, explained about the explosion over Chelyabinsk after he had done considerable homework. His 2003 essay, titled "A Comet's Tale: On the Science of Apocalypse" and which ran in *Harper's* magazine, was an exhaustive journalistic report on end-of-the-world possibilities that could be caused by too-near-Earth asteroids. He asks, "Can you imagine that happening above a major metropolitan area? It would either fill the churches or empty the churches." Bissell went with Steven J. Ostro, a radar astronomer at NASA's Jet Propulsion Laboratory (JPL) in Pasadena who specialized in the asteroid-impact situation, to NASA's Deep Space Network antenna complex at the Goldstone Deep Space Communications

Complex, located in the desert between Los Angeles and Las Vegas, to see for himself where the radar imagery of asteroids and comets comes from. He also interviewed other astronomers at JPL and elsewhere and soon came to understand what they knew about near-Earth asteroids (NEAs) and what they didn't know. He learned about countless tiny asteroids that pelt the atmosphere harmlessly all the time, larger ones in the one-hundred-meter-diameter range whose impact could seriously affect the planet's climate, and the kilometer-or-larger monsters that could bring on doomsday.

"The one-kilometer threshold is important, for asteroids above it are known as 'civilization-enders.' They would do so first by the kinetic energy of their impact, striking with a velocity hitherto unknown in human history"—but well known in dinosaur history. "The typical civilization-ender would be traveling roughly 20 kilometers a second, or 45,000 miles per hour—for visualization's sake, this is more than fifty times faster than your average bullet—producing an impact fireball several times wide that, very briefly, would be as hot as the surface of the sun." If the asteroid hit land, Bissell continued, the smoke from forest fires and other flying debris would shroud Earth in a long night, in a cosmic winter, that would last from three months to six years.[19]

Neil deGrasse Tyson, the effervescent astrophysicist and director of the Hayden Planetarium in New York, offered his own take on the "event" to Clyde Haberman, a veteran reporter for the *New York Times*. "Think of it as a shot across our bow," he said. The shot was a warning about all the potentially dangerous impactors out there, including one called 99942 Apophis. "It's from the Egyptian god of death and destruction," Tyson said. "It was named knowing that it crosses Earth's orbit." Otherwise, "we would have named it something less threatening, like Tiffany or Bambi." Apophis is scheduled to fly uncomfortably close to Earth in 2029, again in 2036, and yet again in 2068, luckily without connecting. But Tyson does not think that luck will be on Earth's side indefinitely. "What is a certainty is that one day Apophis and Earth will collide. So our goal should be, if the survival of our civilization is a concern and a priority, to find a way to deflect it. We know how to do it," he explained, but "there's no funded plan to do so anywhere in the world."[20]

The International Astronomical Union named an asteroid in the Asteroid Belt between Mars and Jupiter after Tyson in 2001. "It's officially called 13123 Tyson, but one can't get too bigheaded about this, given that 13,122 asteroids before mine got named after some other person, place, or thing," he has written in a memoir. "I have nonetheless enjoyed the distinction, and I'm glad, last I checked, it's not headed for Earth."[21]

Indeed, since 13123 Tyson orbits the Sun with countless thousands of other asteroids in the Asteroid Belt, or Main Belt, as it is also called, and is therefore not a dreaded Earth-crosser, so it will never wipe out the human for whom it was named.

While Governor Yurevich was delivering the news about keeping the heat on in Chelyabinsk, Dmitry Medvedev, the prime minister of Russia, confirmed that a meteor had indeed blown up over Siberia. He added that it proved that the "entire planet" is vulnerable to such dangerous occurrences and that a spaceguard system is needed to protect it from similar attacks. Dmitry Rogozin, the deputy prime minister, took it a step further by proposing the creation of an international program that would alert countries to "objects of an extraterrestrial origin" that are potential threats.[22] The international space community readily agrees and is moving in that direction.

And then there were the conspiracy theories, which are as endemic to Russia as vodka, caviar, and the kazatsky. A poll taken by *Noviya Izvestia* (a serious Moscow newspaper and the successor to *Izvestia*, the Moscow daily that, with *Pravda*, were the country's largest and most influential newspapers in the Communist era) showed that about half of its readers accepted the official report that the explosion was caused by a meteor. The other half came up with several explanations, some of them truly bizarre. Some respondents said that they believed the explosion was caused by a secret US weapon test and, by implication, that their government was withholding that fact to save face. Others thought it was an off-course ballistic missile and was therefore top secret. Still others told the newspaper that the violent occurrence high in the sky was a message from God, while some of their countrymen and countrywomen claimed to be convinced that it was a crashing alien spaceship. And there were those who believed Earth had been attacked—zapped—by friendly, protective aliens in the kind of UFO whose platter-shaped predecessors piqued Americans' collective imagination after some were allegedly seen in the 1950s. (The benevolent visitors from Mars and elsewhere helped fill movie theaters that showed *The Day the Earth Stood Still*, *The War of the Worlds*, *It Came from Outer Space*, and other cinematic potboilers.) A fraction of those who were polled confided their suspicion that the object that blew up was an extraterrestrial Trojan horse carrying a space virus that was meant to wipe out Earth.[23]

If Martians or other aliens did indeed want to contaminate the people of Chelyabinsk and then all other Earthlings by blowing up a meteor that carried a deadly virus, they failed miserably. The city and surrounding areas were not stricken by a plague or by any epidemic. The Trojan-horse crowd was in venerable company, though. Isaac Newton, who fled London for the countryside to avoid the plague of 1665, believed that comets that year and the year before had delivered the deadly creatures to Earth. And in their 1979 book, *Diseases from Space*, British astrophysicist Fred Hoyle and Sri Lankan scientist Nalin C. Wickramasinghe unequivocally resurrected the theory of diseases brought to the cradle of humanity by comets, and they did so right up front: "We shall be presenting arguments and facts which support the idea that the viruses and bacteria responsible for the infectious diseases of plants and animals arrive at the Earth from space. Furthermore, we shall argue that apart from their harmful effect, these same viruses and bacteria have been responsible in the past for the origin and evolution of life on the Earth. In our view, all aspects of the basic biochemistry of life come from outside the Earth."[24] It is an appealing theory since spontaneous generation—that life just suddenly appeared naturally—defies logic. That leaves two other possibilities: God created life on Earth, or life came from somewhere else.

A number of astronomers, astrophysicists, and other scientists have taken issue with the cometary-disease theory because they believe it is a gross oversimplification and does not identify a source. "There are a host of objections to life and diseases from comets including no obvious connection between the arrival of large comets in the Earth's neighborhood and worldwide epidemics," Donald K. Yeomans, a JPL astrophysicist, wrote in his book *Comets*. "Influenza viruses, for example, are parasitic and require host cells from specific animals to thrive. Presumably, the necessary animals do not reside in comets," he added with unconcealed scorn. He did theorize that comets may have provided Earth with the basic materials that are necessary for life to form, but he clearly believed that the notion they were arcs that transported life to this world defies reason.[25]

Chelyabinsk's political leaders immediately responded to its brush with catastrophe by ordering a massive cleaning up of the debris, which, as a national news service quoted them claiming, included more than two hundred thousand square meters of broken glass. That estimate, made on the day of the explosion, reflected either extraordinary clairvoyance or, far more likely, unabashed expedience, since the politicians knew that the national government was going to provide the city and the surrounding region with as much money as was required to restore everything that was damaged and to care for the injured. If, in the end, Chelyabinsk was a bit better off than it had been before the meteor appeared, well that was all to the good of the community. That is why, after Governor Yurevich quickly put the damage estimate at more than a billion rubles, several other members of the local government hastened to add that the number would very likely rise. "'Force majeure' circumstances are always a gift to the authorities," Gleb Pavlovsky, a cynical political consultant in Moscow, said of the event, which could be neither anticipated nor controlled, "because you can just write off everything that's stolen."[26] The meteor's explosion, in other words, gave Yurevich and his political cronies an excellent opportunity to enhance their community's net worth—and most likely their own—at the expense of the national government. It would be akin to the US government footing the bill to rebuild a town that suffered property damage because of a tornado or an earthquake and increasing the property value in the process. Twenty-four thousand military personnel and emergency responders were immediately deployed to the afflicted region to assist in the cleanup and to do whatever else was necessary to help the stricken city and towns in the region.[27]

Chelyabinsk's leaders were well aware that the explosion (and the rain of cosmic debris that followed) was a rare occurrence and that it made their city an exceedingly distinctive place; it became a celebrity of sorts among all the communities on Earth. They quickly decided that the meteorite's falling where it did was not just a local event, nor even just a national one, but one that involved the whole world: it was a global event. The city had been randomly selected for an attack by nature as part of Earth's community—it had taken a hit for all of humanity— and it was strongly felt that a monument should be built to commemorate that fact. It is not far-fetched to assume that, consciously or otherwise, some locals recalled that Christ died for others' sins and, in roughly the same vein, Chelyabinsk took a hit for the rest of the planet.

It was quickly decided not only to commemorate the attack

from space but also to profit from it. If the meteor was indeed a rarity—an exotic piece of the sky that had come to Earth—it was reasonable to assume that people with an interest in such things and who were able to pay for them would gladly do so. This was a chance for those who could afford it to actually touch the sky (without having to land on the Moon).

Even as the cleanup was starting, then, a hunt began for pieces of the meteorite that would be marketable. Sasha Zarezina, an eight-year-old girl in the impoverished town of Deputatskoye, which was founded in the 1920s around a collective dairy farm that is long gone, immediately went wading into the snow with other school children in search of fragments, some of which were the size of pebbles, some the size of Ping-Pong balls, and some as big as a fist. First she decided that she would save for her future children the small pebble she found in the snow. But she quickly reconsidered. "I will sell it for 100 million euros," she declared. Larisa V. Briyukova, a forty-three-year-old home-maker, discovered a fist-sized meteorite chunk under a hole in the roof tiles of her woodshed. Then a stranger, one of many who were suddenly driving through the neighborhoods offering stacks of rubles worth hundreds, then thousands, of dollars, showed up and offered her about sixty dollars for it. After some haggling, Ms. Briyukova got the price up to $230. A few hours later, while she was still congratulating herself on having gotten the higher price, another man pulled up, looked at the hole in the roof, and offered her $1,300. "Now I regret selling it," she told one of the foreign journalists who rushed to the region for what was instantly understood to be a very big story. "But then, who knows?" she added after a moment's thought. "The police might have come and taken it away anyway."[28]

While service workers and volunteers worked to pick up smashed glass, repair windows and do light repairs on damaged buildings, many of their friends and neighbors quickly realized that the once-in-a-lifetime occurrence could make money for them; that selling actual pieces of the sky could be profitable. Curious individuals who took time off from work or school that Friday to search for fragments just to see what a chunk of meteorite looked like were quickly joined by friends and neighbors who did it just to turn a ruble. Before nightfall that day, a search for fragments began, and it accelerated over the weekend.

Four days later, the *Los Angeles Times*'s Moscow correspondent reported that the sudden scavenging was being called the "meteorite rush" and that "prices asked for purported pieces of the alien visitor range from $20 to $30,000." His use of the word *purported* showed that he thought it possible, or even likely, that some of those selling what they claimed were parts of the meteorite were in fact hawking counterfeits: that is, Earth rocks that gullible souvenir collectors could be convinced had just come from out of this world. Appropriately, the newspaper ran the story under a headline that announced, "Rubles from Heaven: Russians Scoop up Meteorite Chunks for Sale."[29]

"For sale: a piece of meteorite," proclaimed one ad taken in a newspaper by someone using the name "Yevgeny" who asked $10,000 for a space rock without specifying its size. "Cures cancer, AIDS and prostate. Improves academic performance at school." Another individual claimed in an ad that his or her sample "improves male potency, reduces weight. Price by agreement. Exchange for a car or real estate a possibility." A black porous stone that was about three inches in diameter was put on sale on eBay, and by Tuesday, it had brought in eighty-four bids, the highest being $4,100.[30]

Maxim, a university student from Yemanzhelinsk, a town twenty kilometers south of Chelyabinsk, collected several dozen pieces of what he thought was the meteorite while they were still hot. He said that they were black and hard and looked like porous coal. He kept the largest one, which weighed eighteen grams, gave a dozen smaller ones to a visiting scientist, and put seven pieces that weighed roughly six grams each on sale on the Internet. Maxim (who would not give his last name to protect his privacy) said that he got a dozen calls, including ones from potential buyers in Germany and one from a museum in the United States. He told the *Los Angeles Times* correspondent that he sold one fragment weighing three grams for the equivalent of $150 and wanted the equivalent of $200 for those weighing six grams. No wonder he said that he intended to go back to the "treasure field" to look for more sources of revenue. Conscious of the historical importance of the meteorite, Maxim's fellow townspeople even collected broken glass to give to friends and relatives as souvenirs.[31]

Ms. Briyukova's fears about the police notwithstanding, law enforcement took a decidedly dim view of the craze. "Some people can be easily confused and compelled to buy something they later will be sorry for, as some other people are taking advantage of the situation," Anzhelika Cherkova, a police spokeswoman, said. She told the *Los Angeles Times* correspondent that some "entre-preneurs" might have been trying to make easy money "selling something which has nothing to do with the real meteorite. But even if they are selling the real stuff," she added, "no one knows whether it can pose a health hazard until examined by experts." The fear of contracting something horrible from spaceborne bugs clearly ran through many Russians' minds.[32]

Inevitably, one entrepreneur offered what he called an "Apocalypse tour" around Lake Chebarkul that ended where large chunks of the meteorite were believed to have crashed through the ice, leaving a large hole. Divers searched the bottom of the lake under the hole and found nothing. Andrei Orlov, the mayor of Chebarkul, reportedly wrote on his blog that he had been told that the town of forty thousand "really got lucky." Six days after the remnants of the meteorite plunged into the lake, he called for a citywide brainstorming session on the best way to get the news into travel brochures and books that would attract foreigners and other Russians to the site where the heavens sent an emissary. He and the other town fathers even thought about opening an American-style "Meteor Disneyland"—a theme park that would attract tourists with exhibitions and rides that taught about the dangers lurking in the sky. (They ended the dinosaurs, after all.) The theme park would also celebrate the town for having been the target of an actual impactor and having survived a near miss.[33]

The entrepreneurs were not only in Chelyabinsk and its environs. Some were in America, which abounds in entrepreneurs. An Internet retailer called Zazzle in Redwood City, California, quickly advertised a line of T-shirts that said, "Meteorite Survivor Feb. 15, 2013" and "I Survived the Russian Meteor," available in eighteen colors for $19.40 (plus shipping). There was also an "I Love Meteorites" T-shirt with the usual red heart in place of the word *Love*, a mug with "I eat, sleep and breathe meteorites 24/7" that went for $14.95, a sky rock travel mug at $22.95, an Armageddon poster that cost $21.35, and, perhaps inevitably, a poster that was priced at $12.75 and a sticker that cost $4.95, both of which said, "Kiss My Asteroid." Someone even concocted a Kiss My Asteroid cocktail, with liqueurs and pineapple juice over crushed ice, to sip while giving "that meteor the finger."[34]

As will be seen, catastrophes can be profitable as well as painful.

That Chelyabinsk had been spared a calamity was immediately obvious to its inhabitants, others in the region, all Russians, and foreigners who knew about the incident. But something else good came of it that was not immediately obvious. A common, existential threat brought cooperation and mutual empathy to a troubled industrial community that had long suffered from miserable weather, hard working conditions in factories, and the Mayak and other accidents that resulted in the suffering—or death—of many and made Chelyabinsk the most contaminated place in the world. But the threat that came from space and that did not make distinctions among its victims bonded them into a cohesive and mutually supportive community. They would have remembered that when the danger passed. Common danger from beyond this world can and should compel people everywhere to maintain mutual protection and a common defense of the planet.

2

CHICKEN LITTLE
WAS RIGHT

I t's hard to love an asteroid. Once upon a time, there was a little prince who lived on one that he thought was a star. It was called B612, and it was hardly bigger than he was. "When you look up at the sky at night," he told Antoine de Saint-Exupéry, the French aviator and writer, "since I'll be living on one of them, for you it'll be as if all the stars are laughing. You'll have stars that can laugh!"[1]

Stars may laugh. But with the exception of B612, asteroids are no laughing matter, and neither are their icy cousins, the comets. Every solid body in this Solar System bears scars from collisions with one or the other (or both). Except for His Highness, asteroids in particular have few if any friends, since astronomers and other scientists know that they are exactly what they appear to be: rocks, often laced with iron, moving at such high velocities that they can inflict horrendous damage when they strike.

Comets, which have been imbued with magical beauty in the popular imagination since ancient times—they're the shimmering celebrities of the night sky that have long, curvaceous tails and have, from time immemorial, been taken to be the emissaries of the gods and therefore precede momentous events—are the asteroids' icy counterparts. They, too, carry the potential for death and destruction, but their appearance is deceptive. Asteroids are the obvious, potentially deadly villains of the universe and are generally thought to be an unmitigated menace. But comets have been adored, celebrated, and feared as omens throughout recorded history and undoubtedly long before then.

"Comets, of course, are obvious candidates for the lead in any drama of outer space, because of their spectacular appearance," Robert Shapiro, a retired professor of chemistry at New York University and an authority on the origin of life, has written. "These objects, with their shining heads and long tails, have turned up in the night sky at various times in human history and inevitably made a profound impression. The sight of a comet was taken as a signal that a very important event was about to occur." He quotes Calpurnia in Shakespeare's *Julius Caesar* on the subject: "When beggars die there are no comets seen; The heavens themselves blaze forth the death of princes."[2]

The blazing, long-tailed "messengers" have been credited with carrying complex organic molecules around the universe and, by implication, with bringing life to this planet and very possibly to others as well. Sir Fred Hoyle and Nalin C. Wickramasinghe made that point in *Lifecloud: The Origin of Life in the Universe*, which was published in 1978 and helped to popularize the modern debate on where life as we know it originated. They hypothesized that too many diverse elements of nature would've had to have acted in symphony in a relatively short period of time on this planet for life to have developed spontaneously. "The best explanation therefore of the known facts relating to the origin of life on Earth is that in the early days soft landings of comets brought about the spreading of water and other volatiles over the Earth's surface. Then about four billion years ago life also arrived from a comet that delivered it here. By that time conditions on the Earth had become sufficiently similar to those on the cometary home for life to be able to persist here, probably at first tentatively and then with some assurance as time went on. The long evolution of life on the earth had begun."[3]

Hoyle and Wickramasinghe were so fascinated by comets, and particularly by their potential for being cosmic-life transportation systems, that *Lifecloud* was followed by *Diseases from Space*. Therein, they claimed that comets were also responsible for delivering the viruses and bacteria that inflict infectious diseases on animals and plants on this planet. The first chapter contains the nut graf, as they call it in journalism (or the point of the story):

> We shall be presenting arguments and facts which support the idea that the viruses and bacteria responsible for the infectious diseases of plants and animals arrive at the Earth from space. Furthermore, we shall argue that apart from their harmful effect, these same viruses and bacteria have been responsible in the past for the origin and evolution of life on Earth. In our view, all aspects of the basic biochemistry of life come from outside the Earth.[4]

In the eternal cast of characters in the universe, then, that makes comets both Dr. Jekyll and Mr. Hyde.

Geologist Gene Shoemaker, who invented planetary geology and became the de facto first planetary scientist, may also have thought that stars are a laughing matter. But he certainly knew better than to believe the same about asteroids and comets. He once explained the asteroid threat as he walked around the rim of Meteor Crater near Flagstaff, Arizona, thus putting it in the kind of perspective that is the mark of a great teacher. The crater itself was caused by an impact that was roughly 150 times the force of the atomic bombs that virtually obliterated Hiroshima and killed an estimated sixty-six thousand of its inhabitants. The asteroid or comet that made Meteor Crater would have turned the hearts—or epicenters, as the natural-disaster crowd calls them—of New York, Paris, Moscow, Beijing, Cairo, Nairobi, Melbourne, or any other metropolis packed with people into a vast, smoking ditch filled with grotesquely twisted and smoking rubble and the mangled, bloody remains of millions of people and animals. It has been estimated that a 140-meter-wide rock or chunk of ice—that's about one and a half times the size of the proverbial football field—moving at high velocity could inflict that level of destruction. Shoemaker therefore took the impact hazard out of the realm of the esoteric, which is the way most scientists think about it, and put it where it belongs.

"One thing that makes the comet and asteroid impact hazard so important relative to other hazards is that it is the one hazard that is capable of killing billions of people; of putting at risk our entire civilization," Shoemaker continued. "We can have any number of storms or earthquakes or volcanoes, and they can do terrible damage locally, but they do not put the entire planet at risk the way an impact does." Put another way, the relative difference in destructive capacity between homegrown disasters such Hurricane Katrina—which struck New Orleans and elsewhere at the end of August 2005, killing almost two thousand people and causing almost $82 billion in property damage—or the earthquake that devastated Porto Prince, Haiti, almost five years after Katrina, and a collision with a large asteroid or comet is the difference between the explosion of a conventional bomb and a thermonuclear one.

"It's like being in a hail of bullets going by all the time," Shoemaker added. "They *are* bullets. They're bullets out there in space." For years, Gene Shoemaker and a few of his predecessors and colleagues waged an intellectual war with many other scientists who were just as firmly convinced that craters here on this planet, as well as those on the Moon, Mercury, and elsewhere were caused by erupting volcanoes, not asteroids, comets, and other solid objects that fly around space in all directions at terrific speed. And as creation "scientists" argue that life came to Earth in a highly structured form (though not necessarily on a comet), contrary to what Darwin and the evolutionists who followed him demonstrated, those who believed that all the craters on every planet with a surface and their moons were caused only by volcanoes stubbornly stuck to their theory, evidence to the contrary notwithstanding.[5]

But the evidence of impacts is overwhelming, and the threat is real. Gene Shoemaker's "bullets" are actually tearing around this Solar System by the many thousands, and possibly millions, all the time (though relatively few are potential city-killers, let alone planet-killers). Given the number of planets and moons, collisions with the speeding rocks around them are therefore inevitable. Mercury has so many impact craters that it looks as though it has been bombarded by artillery fire since it came into existence. And that, in effect, is exactly what happened.

Asteroid- and comet-impact craters are almost all over this world. Estimates of the number of large ones varies considerably because, unlike the Moon, Mercury, Mars, and other planets and moons with surfaces, Earth is a living entity and is therefore subject to the growth of vegetation that can cover or disguise craters, deserts and oceans that can hide them, and wind erosion that can practically erase them altogether. Furthermore, establishing size is subjective. The estimated number of large craters—and the definition of size is somewhat subjective—therefore varies from 139 to more than 170. There is overwhelming agreement, however, that the largest crater on the planet that has been discovered so far is the Vredefort crater in South Africa, which is three hundred kilometers wide and was made about two billion years ago. It makes Meteor Crater look like a pothole in comparison.[6]

But it was the impact at what is now Mexico's Yucatan Peninsula, centered at a spot called Chicxulub, that the overwhelming number of experts are convinced finished off the dinosaurs and many other creatures roughly sixty-five million years ago. The explosion played such a decisive role in Earth's history that scientists use it as a dividing event to separate the planet's two distinct geological periods: the Cretaceous and the Tertiary, commonly known as the K-T boundary. That horrendous impact did a great deal more than make the dinosaurs and some other species of animals and vegetation disappear. It profoundly affected the structure of the planet itself by abruptly changing its geology, topography, and climate. But it was by no means the worst hit Earth has taken. The record was established about four and a half billion years ago, when something the size of a small planet crashed into this one with such force that it broke off a huge chunk of the Earth, which went into orbit around itself and became its solitary moon. As is the case with so much of Earth's formative period, how the Moon came to be has also been the subject of widely varying theories.

Like all good scientists, Shoemaker was part Sherlock Holmes, so he searched for and found clues relating to Earth's early history and that of its lone natural satellite. As the great sleuth himself would have said, the clues are often hidden in plain sight and are obvious to the trained eye and open mind. This process can be seen in the dialogue between Sherlock Holmes and Dr. Watson, taken from the 1943 film *Sherlock Holmes and the Secret Weapon* (loosely based on the Sir Arthur Conan Doyle short story "The Adventure of the Dancing Men"). The film is about a scientist who invented a top-secret bombsight and, in order to take the invention for himself, was abducted by Holmes's consummately evil nemesis, Professor James Moriarty. Holmes and Watson attempt to decode a message from the scientist:

> "I'm all in. Can't think anymore," Dr. Watson complained to Holmes after seeing a message written in a jumbled stick figure code by the scientist. "All these letters and figures running through my brain; all twisted round."
> "Twisted round!" Holmes exclaimed.
> "*That's it.* 'Twisted round,' you said. So simple I never thought of it. Reverse the slide. You see that one? It's now identical with the first three names. In other words, all the figures that have number four are written backwards so it reads from right to left, until we reverse the slide, when they read correctly from left to right."

(Elementary, my dear Watson.) Watson was duly awed yet again by his friend and roommate at 221B Baker Street.

That deductive process was used with an equally positive result when Shoemaker and his wife, Carolyn, happened on the town of Nordling while they were on vacation in southern West Germany in 1961. Their adventure was later made into a National Geographic special called *Asteroids: Deadly Impact*, which was filmed for public broadcasting.

Nordling was built in the center of a very large, somewhat-shallow bowl called the Ries Basin. It was there that Shoemaker had his Holmes-like revelation. Looking down at the bowl from its rim, just as he had at Meteor Crater and other large depressions, Shoemaker became convinced that it had in fact been caused by a kilometer-sized or larger rock that struck the site and exploded on impact roughly fifteen million years ago.

Like most such pervasively tidy, red-roofed towns in the region, Nordling has a cathedral named St. George's. And, like the town itself, St. George's dates back to medieval times. The Shoemakers drove into Nordling in an orange-and-blue Volkswagen camper minibus that spring day, parked it, and walked through the town square and up to the front of St. George's. They stopped at the left side of the main entrance. Then he began to look intently at one of the thousands of large stones that had been laboriously chiseled into the shape of rough cubes and cemented in place to make the edifice's walls. The grayish stones caught Shoemaker's eye because tiny parts of them glittered. While scrutinizing one of the stone cubes very closely, he spotted tiny pieces of shimmering silica—primitive glass—embedded in it. He knew that glass is made under high pressure and in extreme heat. Asteroids and comets that slam into the atmosphere do so under immense pressure and become very hot. That's why they glow as they race across the night sky. And the deeper they penetrate into the thickening atmosphere, the hotter they get, so they are hottest at the instant of impact. The rock that made the crater in which Nordling was built, Shoemaker deduced as Holmes would have, was moving so fast and was so hot when it struck that it turned fragments of the rocks it struck into silica. The crater and the silica in its stones that were used to make the cathedral told a very vivid and logical story to Shoemaker. "It was the first impact crater which we could *prove* was an impact crater," he later recalled in triumph. "That just changed the whole ballgame."[7] (Elementary, my dear Carolyn.)

And, Gene Shoemaker added, the next time a really big asteroid or comet strikes the home planet, "It will produce a catastrophe that exceeds all the other catastrophes by a large measure. These things have hit the Earth in the past. They will hit Earth in the future," he added, matter-of-factly. A rock that is more than a kilometer in diameter would create an explosion on impact equivalent to all the world's nuclear weapons going off at the same instant, he said. "Actually," he corrected, "it would be a little bit more energy than that." The eruption of Earth and everything on it would fill the sky with so much debris that it would darken the world for months. It would cause "mass hysteria," Carolyn added.[8]

David H. Levy, a prolific and ubiquitous Canadian with widely varying interests who claims to be "one of the most successful comet discoverers in history" (he spotted eight of them using telescopes in his backyard, according to his resume) is a longtime friend of the Shoemakers who appeared with them on that National Geographic special. He drew a vivid analogy in describing how a human would feel if he or she were near an impact explosion like the one that caused the crater in which Nordling was built. "You would feel as though you were in an oven turned to broil," he said.[9]

The most momentous time in Levy's and the Shoemakers' lives began on March 25, 1993, when Carolyn spotted what looked like a "squashed comet" in a photograph of the space around Jupiter they had taken as part of their routine photos two days earlier at the Mount Palomar Observatory in California.

They were perplexed because, instead of a nucleus trailing a long tail, which is what comets are supposed to look like, this one was bar shaped and seemed to have a series of little tails. Close analysis of the picture under a microscope showed that the "bar" was in fact a string of twenty or twenty-one cometary fragments—a formation of them—bearing down on the giant planet in what was a near-certain collision course. They quickly alerted the astronomical community, and from July 16 to 22, 1994, the three of them and the rest of the world were treated to the first look humans have ever had of a collision between a planet and its cosmic attackers. They watched, almost in disbelief, as the huge chunks of ice successively plowed into Jupiter at a velocity they estimated to be sixty kilometers a second. The largest of the impactors was calculated to be two kilometers in diameter. Each strike left intensely hot gas bubbles and large, dark scars in Jupiter's atmosphere that remained for months. Plumes of gas thousands of kilometers high shot into the atmosphere. It was the first—and so far the only—time when humans have actually gotten to witness an attack on a planet. Levy has written about it, starting with the first of the historic impacts:

> The crash of fragment A was an extraordinary event all by itself. That was the first time that people had witnessed such a collision. Had A been the only impact, we would have rejoiced and studied its effects for years. But in the opera of impact week, A was just the overture. One of the smallest nuclei, A was hitting the planet at a place farther away from Jupiter's daylight side than any of the others, so few scientists—Gene Shoemaker was one—expected to see the plume of material thrown into the atmosphere by its strike.[10]

The opera Levy may have had in mind was Gian Carlo Menotti's *Amahl and the Night Visitors*, whose young hero tells his mother that there is a colossal star over their house, but he is not believed. One after another, the long line of flying icebergs struck Jupiter, each time causing a huge explosion and a gas plume.

What happened next as the astronomers and other scientists watched on monitors the successive impacts belied the widely held belief that science is basically dull and so are its practitioners. "For a moment the group of scientists just sat there, stunned," Levy later reported. "There was a collective gasp. It took a few more seconds before the scientists began to realize what a treasure they had, that in this one picture, all the months of planning had paid off handsomely. 'Oh,' Heidi Hammel exclaimed. 'My God!'"

Levy remarked, "The whole picture was clear now, and the room erupted with cheering and applause. 'We realized that we had something truly spectacular on our hands,' Hal Weaver continued. 'Melissa McGrath ran upstairs to get the champagne that she had bought for the occasion (even though she is the first to admit that she hadn't really expected to see anything like this), and Heidi Hammel popped the cork.'"[11]

New York Times journalist William J. Broad reported, "The Chicken Little crowd, which once drew smiles by suggesting that Earth could be devastated by killer rocks from outer space is suddenly finding its warnings and agenda taken seriously now that Jupiter has taken a beating in recorded history's biggest show of cosmic violence." The story's headline made the point explicitly and accurately: "When Worlds Collide: A Threat to the Earth Is a Joke No Longer."[12]

As is true in most other (pardon) groundbreaking work, the hunt for large, potentially dangerous rocks that have profoundly affected Earth or that could threaten it, perhaps catastrophically, has attracted a variety of people with widely varying opinions.

George Darwin, whose father, Charles, wrote *On the Origin of Species*, one of the towering works of science, was understandably interested in origins, too; though he was interested in the Moon's origins, not *homo sapiens'*. In 1878 he came to the conclusion that the Moon was probably flung off Earth when the planet was mostly liquid and rotating; that is, the two bodies had simply come apart and gone their separate ways. The theory quickly gained popularity not only in the scientific community but also among the general public, where the ancient romantic myth about the Earth goddess giving birth to the Moon goddess was happily resurrected and cherished. Four years later, a geologist named Osmond Fisher embellished the theory when he announced that the Pacific Basin was the birth scar, or depression, that resulted when the Moon separated from her—*her*—mother.[13] If true, it would undoubtedly have been the worst case of a postpartum depression in the history of this world.[14] But no evidence to support that kind of event has ever turned up.

Then there was Thomas Jefferson Jackson See, a supremely fatuous, unscrupulous, arrogant, self-promoting, altogether implausible character whose otherwise credible research and discoveries were almost entirely eclipsed by his blatantly incorrect, but stubbornly held, belief. He was adamant that the Moon was really a full-fledged planet in its own right that came from a place very far away and happened to be passing Earth when it was pulled in by this planet's more powerful gravitational field and was thus forced into a permanent orbit.

See was born in Missouri in mid-February 1866, studied astronomy at the University of Berlin, and, appropriately, died on the Fourth of July, 1962.

He was one of science's most memorable and enigmatic characters: a professional astronomer with a colossally inflated ego who had the manner of a colorful vaudevillian like the ones performing seemingly wondrous feats on stages in music halls while he was doing about the same thing in observatories. See's contributions to astronomy, and to the study of binary stars in particular, were widely recognized and eventually won him mention in the *Encyclopedia Britannica*.[15]

But T. J. J. See was the polar opposite of the popular image of the quiet, self-effacing scientist who puts long, lonely hours of research and discovery above blatantly obvious self-promotion. (That is not to say reputable scientists do not promote themselves. They certainly do. But it is customarily done with apparent modesty and at least the veneer of discretion.) Not so with See. He reveled in the kind of personal and professional self-absorption, bombast, and arrogance that brought scorn and contempt from the staid and usually understated astronomical community. He landed his first job in George Ellery Hale's prestigious department at the University of Chicago, where he was judged not fit for promotion by the distinguished astronomer and was therefore effectively forced to leave. His next stop was the Lowell Observatory at Flagstaff, Arizona, another top-notch institution. But he treated the staff with undisguised contempt for being mere underlings—in effect, his professional servants—and they reacted the way others in their predicament very often do; they found quiet but effective ways to slow their work to the point at which it impeded the astronomy. His departure was therefore inevitable. See's next stop, in 1898, was the US Naval Observatory in Washington, DC, where his carelessness and by then renowned egocentricity again caused angst.

See published *Researches on the Evolution of the Stellar Systems, Vol. II, The Capture Theory of Cosmical Evolution*, a book that ran to more than seven hundred pages and was therefore physically, as well as intellectually, very weighty. In it, he argued that space is not and never was a vacuum. Rather, it was filled with a tangible "resisting medium" that slowed the approaching Moon—which, he pointed out, could not have happened in a vacuum—until, after a great deal of time had passed, it was moving slowly enough to be captured by Earth and pulled into a permanent orbit around the planet. Furthermore, he added, all the planets in the Solar System had been captured by the Sun in precisely the same way. The resisting medium, See explained, was as fundamental to celestial mechanics as Newton's law of gravity.[16]

That the theory was wrong was bad enough. But it was also tainted by a blatant meanness. He described his task, for example, as being necessary to "brush aside the erroneous doctrines heretofore current, as one would the accumulated dust and cobwebs of ages."[17] That gratuitously malicious jab almost undoubtedly left many of his fellow astronomers muttering angry retorts to themselves and their colleagues.

Three years later, in 1913, newspaper publisher and amateur astronomer William Larkin Webb published a saccharine biography of See that once again confounded the astronomical community. *Brief Biography and Popular Account of the Unparalleled Discoveries of T. J. J. See,* as it was called, mirrored his grossly inflated ego. As would be expected, it was swollen with adoring hyperbole. Given See's nature, it was understandable that a rumor quickly started circulating that the self-created superstar of the stars had in fact written it himself.

A reviewer for the *Nation* was so put off by the book's blatant infatuation with its subject—a tone that in other circumstances has been likened to drowning in warm honey—that he or she in effect held up a mirror: "The infant See, we are told, first saw the light on the 393d anniversary of Copernicus's birth . . . [and] showed himself 'every inch a natural philosopher' by speculating on the origins of the sun, moon and stars at the tender age of two, never so much as dreaming that he should grow into a little boy with 'methodical methods,' and one day become 'the greatest astronomer in the world.'"[18] Meanwhile, to his eternal discredit, See also called Albert Einstein a fraud and a plagiarist. Einstein would not lend credence to See by responding, choosing not to sling the mud back and maintaining a dignified silence that added to his stature.[19]

Harold C. Urey, the brilliant, acid-tongued chemist who won the Nobel Prize in 1934 for discovering deuterium, also believed that the Moon was captured by Earth. It was therefore not inherently a satellite of this planet, he reasoned, but rather the fifth of the inner planets that included Mars or the tenth, if the group extended out to Pluto. The latter was eventually demoted to a mere "planetoid" or "dwarf planet" by the International Astronomical Union after astrophysicist Neil deGrasse Tyson, the irrepressibly enthusiastic and animated head of New York's Hayden Planetarium, called for its status to be lowered. He was promptly deluged with hate mail from irate grade-schoolers who identified the newly downgraded (and therefore insulted) member of the Solar System with Walt Disney's beloved pooch of the same name.

So there it was: Darwin the Younger's "fission" theory had it that Earth and its Moon were once a single entity that had come apart on its own in a cosmic split or divorce. See's and Urey's "capture" hypothesis argued that the Moon had once been going its own way but was snagged by Earth and pulled into a permanent embrace. And finally, there was a "coaccretion" theory that maintained both Earth and the Moon were simply created independently when everything else was.[20] The theories were of course considered eminently plausible by their adherents, all of whom believed that theirs was undoubtedly the most likely to have occurred. Collectively, the three explanations seemed to cover all the possibilities.

But nowhere, at least in the astronomical big leagues, was there mention of the fact that the Moon could have been formed when a planet-sized wanderer smashed into Earth with such force that it knocked off a large piece of it, which, over millennia, gradually turned from a jagged chunk of soil, rock, and minerals into a sphere like the other moons and then began circling its "mother." It wasn't until well into the twentieth century that the true level of violence in the universe began to be seen and understood. It was nothing short of phenomenal. As telescopes improved, whole galaxies could be seen to collide, and the Moon's craters, which had been caused far more by impacts than by volcanoes, offered stark proof that Earthlings (and

who knew, maybe Martians, too) live in a universe where violence and destruction occur constantly and are so massive that
they are literally unimaginable ("mind-boggling," the overused
cliché, is appropriate). The mystery of how the Moon came to
be was finally solved when teams of geologists and geophysicists, working independently but with access to some of each
other's research, discovered why the dinosaurs disappeared.

Science is wholly dependent on sharing discoveries, so,
unlike Thomas Jefferson Jackson See, those who investigated
the disappearance of the dinosaurs tended to be generous to
each other and highly professional in describing their monumental collective accomplishment.

The first important clue about this planet's turbulent history,
and therefore about the state of perpetual violence in which it
exists, turned up in Mexico in 1952. Geologists working for
Petróleos Mexicanos, the Mexican national petroleum cartel,
were searching for oil off the Yucatan Peninsula when they found
pieces of hard, dense, crystalline rock instead of the porous, relatively soft, sedimentary rock in which oil is found. Crystalline
rock is called that because it contains crystals. And crystals—as
in glass—are made when minerals are subjected to enormous
pressure and heat. An analysis of the rock samples also showed
that they contained a composition similar to andesite, a fine-
grained rock that is found in volcanic rock.[21] They did not turn
up any oil, and the reason quickly became clear, or at least
the geologists thought it did. Volcanic rock does not contain
oil. It was therefore decided that a long-dormant volcano was
under the water just off the Yucatan coast, beside the village of
Chicxulub.

But Petróleos Mexicanos, ever thinking about the lucra-
tive market for crude oil, would not abandon hope that it
was there somewhere. It therefore sent two geophysicists,
Antonio Camargo Zanoguera, who came from the area, and
an American named Glen Penfield, to press the search in the

late 1970s. They, too, could not find oil. What they did find, however, revolutionized humanity's understanding of its history and, to a considerable extent, its place in the Solar System and the universe beyond. Zanoguera and Penfield discovered the clear outline of an almost-perfect semicircle that measured 180 kilometers across. The perfect geometry intrigued them, so they took a closer look, collecting rock and sediment samples as they went. Since volcanoes do not contain oil, the cartel called off the search. But Zanoguera and Penfield continued to explore the area and soon found the outline of another partial arc on the Yucatan Peninsula itself; on land. The two semicircles came together almost perfectly to form a 180-kilometer-wide circle whose center was near Chicxulub. That circle could have been caused only by an enormous natural "event" of some kind, and there were only two possibilities: a volcanic eruption or an impact. But, except for the material that closely approximated andesite, there was no evidence that the circle had been caused by a volcano. That, in addition to the crystalline rocks, finally started them thinking about the possibility that the great circle had been caused by an asteroid or a comet, which they calculated would have had to have been roughly ten kilometers wide and moving at blistering speed. To use Shoemaker's analogy, it would have been like a bullet fired into a pumpkin.

Walter Alvarez, a geologist at the University of California at Berkeley, became intimately involved in proving that the devastation at Yucatan was the work of an impactor. He has described what he believes would have happened when a very large impactor that he called "Doom" suddenly appeared:

> Doom was coming out of the sky, in the form of an enormous comet or asteroid—we are still not sure which it was. Probably ten kilometers across, traveling tens of kilometers a second, its energy of motion had the destructive capability of a hundred million hydrogen bombs. If an asteroid, it was an inert, crater-scarred rock, dark and sinister, invisible until the last moment before it struck. If a comet,

it was a ball of dirty ice, spewing out gases boiled off by the heat of the Sun, and it announced impending doom with a shimmering head and a brilliant tail splashed across half the sky, illuminating the night, and finally visible even in the daytime as Armageddon approached.[22]

Alvarez is the son of the late Luis Alvarez, a Nobel laureate in physics. He, his father, and two colleagues who were chemists, Frank Asaro and Helen V. Michel, decided to investigate the impact theory in 1980, soon after Zanoguera and Penfield went public with it. Alvarez and his colleagues knew when they began that the element iridium is relatively scarce on the surface of the Earth because it is very dense and much of it, therefore, sank when the planet was forming. But there is a lot of iridium in asteroids and in some meteorites. They found that more of it is concentrated in the sediment at Yucatan than is scattered elsewhere on the planet. In addition, they, like their oil-prospecting predecessors, turned up "shocked quartz granules," which is to say minute amounts of primitive glass. That, again, was clear evidence that the area had once been incredibly hot and under enormous pressure.

"It is very difficult to appreciate the impact that was about to occur, because such an extreme event is far beyond our range of experience—for which we can be most grateful!" Walter Alvarez has written.[23]

One can write down the measures of what happened—an object about 10 km in diameter slammed into the Earth at a velocity of perhaps 30 km/sec. But these measures only acquire meaning when we try to visualize them, or make analogies to help our understanding. How can we imagine a comet 10 km in diameter? Its cross section about matches the city of San Francisco. If it could be placed gently on the surface of the Earth it would stand higher than Mount Everest, which only reaches about 9 km above sea level. Its volume would be comparable to the volume of all the buildings in the entire United States. It was a big rock, or a big ice ball, but not of a scale beyond our comprehension.[24]

What turned it into a cataclysmic weapon was its velocity. The estimated impact velocity of 30 km/sec is 1,000 times faster than the speed of a car on the highway and 150 times faster than a jet airliner.[25]

Walter, his father, Asaro, and Michel wrote a paper on what they discovered for the journal *Science*, which published their findings in its June 6, 1980, issue.[26] That paper played a crucial role in shaping the United States' planetary-defense space program that its authors could not possibly have foreseen.

When that monster struck, it sent so much debris into the atmosphere that the sky was darkened for months, blocking life-giving sunlight and therefore not only finishing off the dinosaurs but also many other species of animals and vegetation. It was, in effect, what would be called a "nuclear winter" during the so-called Cold War that occurred sixty-five million years later, when other dinosaurs, now grossly shrunken but still walking on two legs, reappeared on both sides of what they called an "iron curtain" and threatened to obliterate most of the world with nuclear weapons. The extinction of the great reptiles and more than half of the other species of the time is generally relegated to a time so ancient, so lost in the mist of history, that it is irrelevant. But that is wrong. The huge asteroids and comets are still out there, and the possibility of this planet suffering another catastrophic hit is not only real but also probable. In the words of Robert F. Arentz, a corporate executive who is highly knowledgeable on the subject, "It's not a matter of if; it's a matter of when."[27]

What happened over the Podkamennaya Tunguska River basin in the Central Siberian uplands just after dawn on June 30, 1908, made his point. Eyewitnesses reported that a very intense blue-white streak suddenly appeared in the sky, followed by the sound of a tremendous, thunderous explosion. The blast was so powerful that, sixty kilometers away, it knocked people off their feet. "I was sitting on the porch of the house at the trading station, looking north," a man who lived close by later reported. "Suddenly in the sky north . . . the sky was split in two, and high above the forest the whole northern part of the sky appeared covered with fire. I felt a great heat, as if my shirt had caught fire. . . . At that moment there was a bang in the sky, and a mighty crash. . . . I was thrown twenty feet from the porch and lost consciousness for a moment. . . . The crash was followed by a noise like stones falling from the sky, or guns firing. The earth trembled. . . . At the moment when the sky opened, a hot wind, as if from a cannon, blew past the huts from the north. It damaged the onion plants. Later, we found that many panes in the windows had been blown out and the iron hasp in the barn door had been broken."[28]

And this from another survivor: "The ground shook and incredibly prolonged roaring was heard. Everything round about was shrouded in smoke and fog from burning, falling trees. Eventually, the noise died away and the wind dropped, but the forest went on burning. Many reindeer rushed away and were lost."[29]

Nineteen years later, in 1927, the first group of scientists reached the desolate area and were stunned by what they found. Most of the trees within thirty kilometers of where the midair explosion had occurred were down and heavily charred. And the fact that all of them were in a circle that pointed away from the center, plus the absence of a crater, left little doubt as to what had happened. Soviet scientists concluded that a stony meteorite measuring about one hundred meters in diameter had exploded as it penetrated the thickening atmosphere.[30] Scientists have disagreed for years over whether the explosive near miss over Tunguska was a comet or a meteoroid. There was no disagreement within the community, however, about its having been part of a large population of near-Earth objects (NEOs), some of which are potentially very dangerous.

David Morrison is the director of the Carl Sagan Center for

the Study of Life in the Universe and a senior scientist at NASA's Ames Research Center in California. He has been researching NEOs for most of his professional life and is an acknowledged expert on the subject. In 2000, he began sending out via the Internet frequent bulletins that were loaded with NEO developments. They were and still are simply called *NEO News*, and they live up to their name.

"We have a long-standing research program to understand more about comets and asteroids, including those that come close to the Earth," Morrison has explained. "While the research program is motivated primarily by a desire to understand the scientific aspects of these small bodies, it is also designed in part to provide the information that will be essential for planning a defense program against hazardous impacts." Like his colleagues in NASA and elsewhere, Morrison is not an alarmist and thinks about the situation objectively. "I can tell you with confidence that for the ten percent of the big ones that have been discovered, there is no danger. But I can tell you nothing about the ninety percent that we have not discovered. So yes, we understand the nature of the risk, but we have not taken any concrete efforts to protect ourselves or even to look to see if there's anything headed our way."[31]

Morrison continues, "Cosmic impacts are highly infrequent, and the largest (mass extinction level) events have characteristic time-scales of tens of millions of years. Even the smaller localized events have low probability relative to other more familiar natural hazards such as earthquakes, tsunami waves, and severe storms. Until astronomers began to survey for potential impactors, the risk was perceived as random, and little, if any, warning could be expected. From the perspective of an elected official, the chances of having to deal with such a catastrophe within a term of office are extremely low, whether we are discussing local or global events. Yet the potential exists for an impact catastrophe at any time, in any country, with little or no warning."[32]

"For some people, meteorites are trophies, to be cherished and displayed," Neil deGrasse Tyson wrote in his autobiography, *The Sky Is Not the Limit.*

> For me, they are also harbingers of doom and disaster. Consider that the slowest speed a large asteroid can impact Earth is about six or seven miles per second. Imagine getting hit by my overpriced objet d'art moving that fast. You would be squashed like a bug. Imagine one the size of a beach ball. It would obliterate a four-bedroom home. Imagine one a few miles across. It would alter Earth's ecosystem and render extinct the majority of Earth's land species. That's what meteorites mean to me, and it's what they should mean to you because the chances that both of our tombstones will read "killed by asteroid" are about the same for "killed in an airplane crash."
>
> About two dozen people have been killed by falling asteroids in the past four hundred years, but thousands have died in crashes during the relatively brief history of passenger air travel. The impact record shows that by the end of 10 million years, when the sum of all airplane crashes has killed a billion people (assuming a conservative death-by-airplane rate of a hundred per year), an asteroid is likely to have hit Earth with enough energy to kill a billion people. What confuses the interpretation of your chances of death is that while airplanes kill people a few at a time, our asteroid might not kill anybody for millions of years. But when it hits it will take out hundreds of millions of people instantaneously and many more hundreds of millions in the wake of global climatic upheaval.[33]

The likelihood of that happening, or at least of an asteroid colliding with Earth irrespective of the death and destruction it could cause, was put on a scale in 1999 by Richard P. Binzel, a professor of earth, atmospheric, and planetary sciences at MIT. He called it the Torino Impact Hazard Scale in honor of the Torino Observatory in Turin, Italy, which had done advanced research of asteroids for two decades. Binzel first described the scale at the United Nations International Conference on Near-Earth Objects that was held in April 1995. It takes into account the NEO's size, speed, and direction and assigns it a number from zero (which means that there is virtually no chance of impact or damage) to ten (which means that there will almost certainly be a catastrophic collision). The number is calculated by the astronomers who track the asteroid and is then announced to the scientific community and to the public. The idea was to come up with a way of categorizing threats that is consistent, and it worked.

By the start of the last decade of the twentieth century, Congress had become sufficiently concerned about the impact threat, so it mandated NASA to locate within ten years 90 percent of NEOs with diameters of a kilometer or more. Then, in 2005, Congressman George E. Brown of California, whose constituents included several aerospace companies, led a campaign that expanded the number of potentially threatening intruders the space agency had to locate and catalog to 90 percent of those that measured 140 meters or larger by 2020 (as in perfect vision). But the Obama administration did not ask for funds to complete that task, and the concern that was dramatically proclaimed in Congress did not extend to authorizing enough of an appropriation to do the necessary work. As has been ruefully noted for a very long time, the members of the House of Representatives seem to have attention spans that are limited to two-year election cycles. The space agency therefore had to try to carry out its mandate with restricted funds. It was like a modern farmer being ordered to produce a bumper crop using only a plow pulled by a mule.

But two usable crops—that is, studies—were harvested. The first, titled *Asteroid and Comet Impact Hazards* and conducted by the Spaceguard Survey Report, was released in 1992. The second, *Report of the NEO Survey Working Group*, came out three years later. The Spaceguard Survey Report owes its name to Arthur C. Clarke, who invented the term and used it as the title of chapter 1 in his *Rendezvous with Rama*: "SPACEGUARD." So too does the concept behind the report owe itself to Clarke: it made the obvious point that defense against asteroids and comets depends on understanding their nature—that is, their size, their composition, their velocity, their location, and the direction in which they are moving. The Spaceguard Survey Report ranked the potential impactors' destructive capacity according to size, which translated to "kinetic energy." Smaller ones—those with less kinetic energy—approach Earth more frequently than their larger counterparts but inflict little or no damage. The bigger ones come in less frequently, but when they do, they can inflict severe damage. And coastal populations are at greater risk than inland populations because of tsunamis. "Persons living in coastal regions," the report warned, "run a risk from impact-generated tsunamis as much as two orders of magnitude greater than that from land impacts."[34]

As the Spaceguard Survey Report was accepted and began to take hold within the space community, so did the effect of an important paper that was published in February 2001 that brought the planetary-defense situation around full circle. *The Comet/Asteroid Impact Hazard: A Systems Approach* was written by Clark R. Chapman, Daniel D. Durda, and Robert E. Gold, three leading space scientists. Chapman and Durda were in the Department of Space Studies at the prestigious Southwest Research Institute in Boulder, Colorado, and Gold was in the equally prestigious Space Engineering and Technology Branch of the Applied Physics Laboratory at Johns Hopkins University in Laurel, Maryland. They maintained that the danger of a serious impact was potentially so great that an integrated approach to the problem across the scientific, technological, and public-policy sectors was fully justified. While the emphasis at that point had been on astronomers spotting potential impactors and finding ways to deflect them, the authors contended that little or no thought had been given to connecting the astronomers with the military and civilian agencies that would be responsible for pushing approaching NEOs off course; to planning other kinds of mitigation and dealing with the repercussions of predicting an impact (mass civil panic, for example); and to the need for an informed, international effort to actively plan a mitigation strategy that would replace what they derided as "unbalanced, haphazard responses." *Mitigation* in the NEO community means "to stop the threat." "For example," the authors contended, "we believe it is appropriate, in the United States, that the National Research Council develop a technical assessment of the impact hazard that could serve as a basis for developing a broader consensus among the public, policy officials, and government agencies about how to proceed. The dinosaurs could not evaluate and mitigate the natural forces that exterminated them, but human beings have the intelligence to do so."[35] Then, poignantly, they took note of the seed that had been planted two decades earlier:

> The scientific community began to understand the implications for life on Earth of errant small bodies in the inner Solar System in 1980 when Nobel Laureate Luis Alvarez and his colleagues published an epochal paper in *Science* (Alvarez et al. 1980) advocating asteroid impact as the cause of the great mass extinction 65 million years ago that led to the proliferation of mammal species based on an extraordinary discovery they made at the Chicxulub impact crater underwater off Yucatan on Mexico's gulf coast.[36]

The National Research Council, which is part of the prestigious National Academy of Sciences, was paying attention, and so was the rest of the international space-science community. In October 2002, Chapman met with former astronauts Russell L. "Rusty" Schweickart and Ed Lu, as well as Piet Hut, an astrophysicist at Princeton's Institute for Advanced Study who specialized in planetary dynamics (specifically in preventing asteroid impacts). They gathered at the Johnson Space Center and hatched the B612 Foundation, which was named in honor of the little prince who lived on the asteroid. The foundation's purpose was to protect Earth from asteroid strikes by funding the development of ways to deflect them.

In February 2000, a far-ranging space probe called the Near-Earth Asteroid Rendezvous Shoemaker (NEAR Shoemaker) was ordered to fly in close formation with a thirty-three-kilometer-long, thirteen-kilometer-wide, potato-shaped asteroid named 433 Eros for a year, collect data as it went, and then land on it. Eros, as in *erotic*, is the god of sex and making love in Greek mythology. NEAR Shoemaker's controllers, who, like many in the space program, had a sublime sense of wit and irony, ordered it to mount Eros on February 12, 2001. It was therefore on Eros two days later, which was Valentine's Day.

The next step in planetary protection was a big one, and it happened at the Hyatt Regency Hotel in Garden Grove, California, from February 23 to 26, 2004, when the National Research Council held the first international Planetary Defense Conference. The conference was dedicated to "Protecting the Earth from Asteroids"[37] and was precisely what Chapman, Durda, and Gold had called for three years earlier in *The Comet/Asteroid Impact Hazard: A Systems Approach*.

Eighty-one papers were presented at that meeting on topics such as "Order-of-Magnitude Analyses of Planetary Defense Problems," "Orbit Determination for Long-Period Comets on Earth-Impacting Trajectories," "The Impact Imperative—A Space Infrastructure Enabling a Multi-tiered Earth Defense," and "The Mechanics of Moving Asteroids." In addition to heavy participation by Americans, scientists from Russia, Italy, Ukraine, Bulgaria, Spain, South Korea, France, Germany, and India came to make presentations.

At about the same time, and unrelated to the meeting, David Morrison began sending out *NEO News* on the Internet, and it subtly but effectively helped to bring the NEO community together. And a galaxy of astronomers and others who were interested in the impact threat began to coalesce as a separate group.

The reach of Morrison and several other astronomers soon extended to Europe. On March 26, 1996, the Spaceguard

Foundation was established in Rome with the declared intention of protecting "the Earth environment against the bombardment of objects of the solar system (comets and asteroids)."[38] Its members were mostly European, of course, but included a Japanese scientist on its board of directors. And the foundation's two trustees were definite attention-getters. The first is Fred Whipple, a very well-known astronomer. The second, none other than Arthur C. Clarke, who was billed as the author of "several famous science fiction novels, including . . . *2001: A Space Odyssey* (1968); in *Rendezvous with Rama* (1973) he described the effects of the collision of an asteroid with the Earth and the settlement of a *Spaceguard* organization for the protection of the Earth against such events."[39]

Then the European Space Agency weighed in. In April 2006, it announced that it planned to "slam" an impactor probe into an asteroid in a mission it wryly called Don Quijote. The mission was to involve two spacecraft: one named Sancho that would orbit the asteroid and study it for several months, and a second, Hidalgo, that would collide with it. Then Sancho was to return to the asteroid to assess the damage and report home.

The mission was originally scheduled for 2011; then it was postponed to 2015. And then, in late December 2009, Anatoly Perminov, the head of Roskosmos (the Russian space agency), declared that he was considering inviting the international community to send out a robot to meet a large asteroid named Apophis that is headed in this general direction and to nudge it off course before it makes its first pass by Earth in 2029. He was referring to 99942 Apophis. The huge rock takes its name from an evil ancient-Egyptian serpent that dwelled in eternal darkness in the center of Earth. No professional astronomer seriously believes that Apophis is going to hit this planet the first time around, but there is some concern that it, too, will be ensnared by Earth's gravity and swing into an orbit that becomes increasingly closer until it impacts.[40]

The possibility of Earth attracting a potential impactor came to mind on June 6, 2002, when the Eastern Mediterranean event took place. Yet another meteor exploded in the sky between Libya and Crete, without warning and with the power of a small atomic bomb. It happened during the 2001–2002 India-Pakistan confrontation, and, in the opinion of Gen. Simon P. Worden, vice director of operations for the US Space Command, had it occurred three hours later, it could have caused a nuclear war between the two enemies.[41] Either side could have thought the explosion was caused by a ballistic missile that had been launched by the other side, blowing up prematurely, or that it was meant to knock out the other side's communication capability as a prelude to a nuclear war. Seen in that light, the name *impactor* takes on two distinct and dangerous meanings. They are, of course, physically dangerous. But to the extent that they can trigger war or other forms of manmade violence, they are abidingly dangerous politically as well.

3

KNOW THINE ENEMY

"**S**o it is said that if you know others and know yourself, you will not be imperiled in a hundred battles," Chinese warrior-philosopher Sun Tzu wrote more than two thousand years ago in his classic work, *The Art of War*.[1] "Others" is often translated as "enemies," which is precisely what he meant. "Know thine enemy" is a widely used variation on the theme. And since there are an infinite number of others—enemies—in this world, the advice is universally applicable to individuals and to the multifarious relations between nations.

There are also infinite enemies out of this world, and knowing all about them—what they are, where they are, and where they are going—could make the difference between this planet's safety and survival and its either suffering a severe wound or being annihilated altogether. The early Greeks and Romans saw that rocks fell from the sky and pondered its meaning. Aristotle was convinced that they were first lifted off Earth's surface by strong winds and were then thrown back. He and Pliny the Elder are thought to have both written about an impact that occurred in 467 BCE, when a meteorite fell on Aegospotami in Thrace, on the European side of the Dardanelles. And a comet was seen the same year.[2]

"Thunderstones" were held in awe and venerated as thunderbolt weapons that were hurled at this world by angry gods. In his book *Natural History*, Pliny wrote that a comet appeared in 44 BCE just after Julius Caesar was assassinated and during athletic games he had sponsored. Pliny reported that it was seen everywhere as a bright star in the north for seven days. It was widely believed that the comet was the soul of Caesar on its way to the sanctum, where the immortal gods lived. The emblem of a star was therefore added to a bust of the slain emperor that was dedicated in the forum.[3]

Comets soaring across the night sky, trailing their long luminescent tails, excite us because we take them to be elegant visitors from the infinite, black universe—space—that engulfs us. (And since *vacuum* is defined as a place absolutely devoid of matter, it does not apply to space, which is full of stars, planets, moons, asteroids, comets, and all manner of cosmic debris that are the size of pebbles and grains of sand.) But that take is new and is the product of modern astronomy. For centuries, comets were taken to be fiery messengers, flaming arrows, that were shot at Earth from angry gods. Unlike sunrise, sunset, the stages of the Moon, and other celestial occurrences, comets were unpredictable, and that caused widespread fear and superstition among people who found security in the predictability of nature. The sudden appearance of a comet was taken to be a harbinger of doom; a message from their god, who was angry and warning that a war, a natural disaster, a plague, a famine, or some other catastrophe was imminent. The Babylonian *Epic of Gilgamesh*, the most ancient surviving myth, warned that a comet would bring fire, brimstone, and a flood. The Sibylline Oracles, who were Roman prophets, referred to comets as a "great conflagration from the sky, falling to earth."[4]

Chinese astronomers as early as the Han Dynasty kept extensive records on the appearances, paths, and disappearances of hundreds of comets and associated different shapes with specific disasters.[5]

The beginning of serious, modern astronomy that started when Copernicus decided that the Sun, not Earth, was the center of this planetary system took a colossal leap forward when Galileo began peering through that first telescope—a long, narrow, wooden tube that was covered with paper and had a small lens at either end—in 1609. Myth was replaced by hard reality, and a revolution started that slowly but surely gained momentum as more of the devices were built and used by men who were enthralled to be able to finally get much closer views of the heavens; to, in effect, get off here and go out there. One of them was Johannes Kepler, a German astronomer and astrologer (they were often considered indivisible) who used his celestial magnifier to study planets and then published the three laws that describe their orbits and that bear his name.

One night in 1682, seventy-three years after Galileo trained that crude cylinder on the Moon's battered surface, a twenty-six-year-old Englishman named Edmond Halley who had a deep interest in natural science, including oceanography and meteorology, in addition to mathematics, noticed a comet flash across the sky and instinctively began plotting its positions when it reappeared on successive nights. In August 1684, he visited the University of Cambridge to see an obscure professor named Isaac Newton. Halley told Newton about the frustration he and some friends had had in trying to understand the force that kept the planets traveling around the Sun and that grew weaker with distance. Objects moved faster when they were closer to the Sun, Halley explained, and slower when they were farther away.

"I have calculated it," Newton told Halley. "I worked it out seventeen years ago. I wrote the full solution down five years ago. It's around here someplace," the genius told his visitor. Then history's most famous absent-minded professor shuffled through some papers.

"You can't be serious," said Halley. "You explained the force that controls the planets and you never told anybody about it?"

"It explains the Moon in orbit around the Earth and the tides also. And how dropped objects fall to the ground. Let's see. I think I put it in this drawer. No, not here, either," the man who conceived the law of universal gravitation told his awed visitor.[6]

It was eventually calculated that the comet Halley had noticed and then took pains to plot returned every seventy-six years and, because of its brightness, it (like other comets) was seen, tracked, and studied. In keeping with tradition, the comet Halley studied so intently was named after him. And sure enough, as predicted, it returned on Christmas night in 1758, sixteen years after the man who immortalized it had passed on. It last crossed Earth's path in February 1986, and astronomers were able to see that it is made of water ice, carbon monoxide, carbon dioxide, ammonia, methane, various other chemicals, and some iron. It was appreciated by Earthlings who understood that they were witnessing what for almost all of them was a once-in-a-lifetime event that was savored. It is due back in 2062.

Astronomers and space scientists certainly appreciate comets for their appearance, but these are the people on the most-wanted list of those responsible for "mitigating" comets' potential to do terrible damage on impact; that is, they are in charge of finding and stopping a comet before a collision that is bound to be terrible. That is why Clark Chapman, Daniel Durda, and Robert Gold wrote *The Comet/Asteroid Impact Hazard: A Systems Approach*. Their concern was shared by many in the space community, who organized the groups that are described in chapter 5: "The Other Salvation Army." Scientists, including Gene Shoemaker, Clark Chapman, Donald Yeomans, Steven J. Ostro, Jon D. Giorgini, and David Morrison (a stalwart among the near-Earth-object fraternity who is at the NASA Ames Research Center in Northern California), have found comets so interesting that they specialized in studying them. Yeomans, as noted, was at NASA's Jet Propulsion Laboratory (JPL), where he headed the Near-Earth Object Program Office; so were Ostro and Giorgini. JPL ran most of NASA's Solar

System Exploration Program in the 1960s and '70s, most notably Voyager 1's encounter with Jupiter, Saturn, and several of their moons and Voyager 2's phenomenal "Grand Tour" of Jupiter, Saturn, Uranus, Neptune, and many of their moons. It is important to note that, in addition to sending home a veritable library of new scientific information about those distant worlds—that Jupiter's moon Europa has an active volcano; that the planet's fabled rings are made of an infinite number of rocks of all sizes and pebbles; and that Neptune has rings, too—the stream of imagery that came back to JPL from every close encounter with a solid body showed impact craters that were made by collisions with asteroids of all shapes and sizes. It was the pockmarked moonscape, the eternal target of sheer violence, repeated infinitely. The wandering rocks themselves, which were known to be an important cause of destruction in the Solar System, were studied with obsessive interest by many astronomers and planetary scientists, who then produced hundreds of scholarly papers and books, such as *Comet/Asteroid Impacts and Human Society: An Interdisciplinary Approach* edited by Peter T. Bobrowsky and Hans Rickman and *Asteroid: Earth Destroyer or New Frontier?* by Patricia Barnes-Svarney.

The ubiquitous (and mellifluous) astronomer Carl Sagan, who was unfairly denounced by some members of scienceworld for being a dreaded "popularizer," wrote *Comet, Revised* with Ann Druyan, which was published in 1997. Two chapters were devoted to asteroids and comets hitting Earth and other targets and featured a ride on one (possibly sending a message to Michael Bay, Jerry Bruckheimer, Bruce Willis, and the rest of the cast and crew of *Armageddon*, which was released the following year). That was followed by *Cosmos*, a book that became the television series that made Sagan famous. Yeomans and three colleagues published *Mitigation of Hazardous Comets and Asteroids* in 2004, and he wrote *Near-Earth Objects: Finding Them before They Find Us*, a serious, readable primer on the subject (and on space in general), almost a decade later. And Duncan Steel, an Australian astronomer, weighed in with *Rogue Asteroids and Doomsday Comets: The Search for the Million Megaton Menace That Threatens Life on Earth*.

Not to be left out, a Russian astronomer and one from Belarus weighed in with news about a huge comet that came out of the Oort Cloud, which is believed to envelope this Solar System and sends in every comet. This large comet, they predicted, will visit the neighborhood but not pose a threat to Earth. It is variously called Comet C/2012 S1, Comet ISON (for the International Scientific Optical Network, one of whose telescopes was trained on it), and Nevski-Novichonok for the two men who discovered it.

On at least one occasion, the dramatic specter of a comet streaking across the sky like a shot from God's sling caused a bizarre tragedy. The comet was Hale-Bopp, which was discovered in July 1995 by two independent American sky watchers after whom it was named: Alan Hale and Thomas Bopp. It caught the attention of the public when it became a spectacular sight as it passed Earth at distance of only seventy-six thousand kilometers in March 1997 and was acclaimed the Great Comet of 1997 by millions of people around the world who could see the spectacular show it put on just by looking at it without telescopes.

Thirty-nine of those who looked up lived in San Diego and belonged to a doomsday religious cult called Heaven's Gate, which was founded in the early 1970s by a recovering heart-attack victim named Marshall Applewhite and his nurse, Bonnie Nettles. Both Applewhite and Nettles believed that Earth was about to be recycled, which is to say wiped clean of the creatures that lived on it and then renewed. The two claimed to be from "somewhere else" in the literal, not figurative sense— extraterrestrials—and insisted that the only way to avoid the apocalypse and find salvation at the Next Level (of existence) was to shed every attachment to the planet, starting with ending all relationships with others, including family and friends, and forgoing individuality, jobs, possessions, money, and sexuality so their spirits would not be encumbered by earthly baggage for the voyage to the Next Level. The cult rented a 9,200-square-foot mansion in a gated, up-scale community and spent $10,000 on alien-abduction insurance to cover up to fifty people. Hale-Bopp became visible to the naked eye in May 1996, and the leader of Heaven's Gate was soon telling his devoted followers that the comet was being followed by a UFO that would transport their spirits to the Next Level, but only if they shed the excess baggage they called their bodies. The followers did that by swallowing phenobarbital, cyanide, and arsenic and washing them down with vodka over the course of three successive days, with one group carefully laying the "embarked" deceased in their own bunk beds, their faces and torsos covered with a square purple cloth, before it took its turn with the poison. All thirty-nine decomposing corpses were dressed in identical black shirts and sweat pants, wore armband patches that said "Heaven's Gate Away Team," and carried a five-dollar bill and three quarters, which they apparently believed was the toll that had to be paid to get to the Next Level.[7] Such is the often-perverse romance of comets, however dangerous they can be.

Monster asteroids are far more dangerous than comets because the latter are relatively soft and there are far fewer of them. The little prince may have loved the asteroid he lived on, but the rocks that roam inside the Main Belt and randomly cross Earth's path do not inspire affection from most of the planet's inhabitants, especially those who know firsthand what they do as meteorites, meteoroids, or bolides, as they are called when they get closest to Earth. (Bolides are loosely defined as fireballs.) Ask anyone in Chelyabinsk. One colossal object—a projectile roughly the size of Mars—is thought to have smashed into Earth during its formative stage with such force—a "giant impact"—

that it knocked off a chunk that became the Moon. The author of a book on that subject gave it a title that is as vivid as it is imaginative: *The Big Splat*.[8] Earthlings don't like and, indeed, fear asteroids because it is they, not comets, that have had the overwhelming number of collisions with this planet.

Earth's oceans and shifting terrain hide, obscure, or erase many impact sites. Yet more than 160 remain on all seven continents, many of them chillingly massive. The Sudbury crater in Ontario, Canada, is slightly more 130 kilometers across, and the Vredefort ring in South Africa is 190 kilometers in diameter and looks like an immense flat canyon when observed from high in the air or from space. The Acraman crater in Lake Acraman, South Australia, is a scar that was made about 580 million years ago and measures 90 kilometers across, while the Woodleigh crater in Western Australia was made roughly 364 million years ago. The Manicouagan crater in Quebec, which is 100 kilometers wide, resulted from an impact made 215 million years ago. And a crater was discovered in Chesapeake Bay, 201 kilometers from the nation's capital, in the early 1980s. Its estimated age is 35 million years, so it cannot be blamed on an incompetent administration, Republican or Democrat.[9]

There are innumerable examples all over Target Earth. The Hoba meteorite is a sixty-ton slab of iron that slammed into what is now Namibia about eighty thousand years ago, and since no one in that country seems to have the inclination or resources to move it, it has become a national monument to be appreciated and studied. The Willamette meteorite, which is a ten-foot-tall mass of pitted iron that weighs around fifteen tons, is thought to be the worldly remains of the iron core of a planet that crashed into what is now this one billions of years ago. It was discovered on Native American land in 1902, the local indigene's pleas to leave it alone because they believed it had healing power were ignored, and it was unceremoniously taken away and wound up on display at the American Museum of Natural History in New York. Another one, ALH 84001, is one of the most intriguing artifacts from out of this world. Its dull name notwithstanding, it comes from Mars, was discovered in Antarctica in 1984, and, after scientists saw on close examination, contains the fossilized remains of what may have been very small bacteria. If they really were living microscopic organisms, it would be further proof that Martians, however tiny, have indeed lived on the Red Planet.

Bombardment by near-Earth objects (NEOs) is widely associated by the public with the demise of the dinosaurs and, therefore, with prehistoric and ancient "history." But Tunguska and Chelyabinsk were dramatic demonstrations that the attacks are still underway, and those "events" were in fact only two among thousands that have occurred since records of the rain of rocks and ice started being kept in the eighteenth century.

Like every other body in the Solar System, Earth is under constant bombardment from objects ranging in size from tiny grains of sand to enormously large rocks, many of which are laced with iron. Most of them come from the Asteroid Belt between Mars and Jupiter and are the remnants of a planet that failed to form because of Jupiter's and Saturn's gravitational pulls (and with a little help from Mars). The asteroids in the belt collide with each other constantly and break into fragments that head in all directions, including this one, at velocities that can reach fifty thousand miles an hour. The small rocks range from the size of a beer can to a Winnebago. Those in the medium category start at the size of a two-car garage and can be as large as a small building in the average town. The big ones would sit in a professional football stadium the way a hard-boiled egg would sit in a porcelain cup. And the really big ones start at a kilometer across and go upward to perhaps six to twelve kilometers. The good news is that the size of the asteroids is inversely proportional to their number, meaning that there are relatively few of the really big ones in the kilometer-

or-larger class—also known as the Doomsday rocks. That means the neighborhood is fairly infested with thousands that could demolish downtown Peoria, Rangoon, or Nairobi, but the ones that are large enough to end civilization show up only about once every one hundred years. Some are round, some look like dog bones, and some like Idaho potatoes. Their composition varies, as well. The overwhelming majority of Gene Shoemaker's bullets are solid rocks, but there are also relatively soft ones that are of a substance like sand held together with glue, while still others are made mostly of iron.

The bullets speed around the planet in all directions and at all altitudes all the time. That is why there were major impacts on or explosions directly over Brazil, Canada, the United States (Michigan), Italy, Spain, the Indian Ocean, the Eastern Mediterranean, and Indonesia between 1930 and 2009.[10] The explosion over Brazil happened above the Amazon in the northwestern part of the country, just after sunrise on August 13, 1930, and left physical and emotional scars that are still there. As women began to wash clothes in the Curuçá River and their men fished or tapped rubber trees, the Sun turned bloodred. Then the region was enveloped in darkness. Suddenly there were three very loud whistling sounds, followed by three loud explosions in rapid succession. The blasts turned the jungle into an inferno that burned for several months, which sent many villagers in the region, fearing imminent death, fleeing with as many belongings as they could carry. But others stayed in their huts in an area of several hundred square kilometers. Terrified children tried to hide in corners while some of their parents listened to what sounded like artillery shells and looked at the sky to see what was happening. The ground trembled as it would in an earthquake. Five days later, an Italian monk named Father Fedele d'Alviano began his annual missionary visit to the area and found people who were still shaken by the event. He tried to reassure them that they had suffered from meteorites coming from the sky, not from the wrath of God. It was subsequently called a possible "Brazilian twin" to the Tunguska explosion.[11]

And at least two impacts on the homes of individual Americans were offbeat enough to be reported by the news media as just plain quirky. On November 30, 1954, Mrs. Hodges was severely bruised when a five-kilogram rock came crashing through her roof in Sylacauga, Alabama, and hit her while she was napping in her living room. She got off with a severe bruise. (Mrs. Hodges happened to live across the street from—you can't make it up, as some in the news business say—the Comet Drive-In, a movie theater.) And thousands saw another rock as it sped over Kentucky and headed northeast before making a sonic boom and breaking into more than seventy fragments on October 9, 1992. One football-sized fragment that weighed nearly two pounds landed on Michelle Knapp's Chevrolet Malibu that was parked in her driveway in Peekskill, New York, that day. Its trunk was scrunched, but the $69,000 that a consortium of three dealers paid her for it more than took care of the repairs.[12] As the dealers in Chelyabinsk were to learn twenty-one years later, souvenirs from heaven are valuable collector's items.

Although some members of Congress undoubtedly read about what happened to Hodges and Knapp, they didn't need those incidents to conclude that their planet was and still is under continuous attack by large and small rocks coming from space and that the dinosaurs were almost undoubtedly terminated by a really big one. Those isolated events would have gone virtually unnoticed by Congress, and certainly by the international community, since, by their nature, they were of no real consequence to the rest of the world. Except that just about everyone who was paying attention knew that some of the rocks out there were a lot bigger than what came down on Sylacauga and Peekskill. Gene Shoemaker had not yet participated in the film *Asteroids: Deadly Impact* for National Geographic Video, but it was generally understood that a collision with a large-enough impactor could inflict more damage than a nuclear war and even end their planet's existence. The explosion of a 450-ton meteor twelve miles above Chicora, Pennsylvania, on June 24, 1938, and others over British Columbia in March 1967, Lake Huron in September 1966, and Greenland in December 1997, made the point. As British poet and critic Samuel Johnson said, "Depend upon it, sir, when a man knows he is to be hanged in a fortnight, it concentrates his mind wonderfully."

When the comet Shoemaker-Levy 9 slammed into Jupiter in July 1994, it was extensively covered by the news media and concentrated the legislators' and the rest of the world's collective mind. The images of the successive impacts and the inevitable references to what any one of them would have done to Earth drove home the need to find out what, precisely, the nature of the threat was and then come up with an effective defense, or "mitigation," as the community calls it. Every other conceivable problem was dwarfed by the prospect of Doomsday: by the possibility, however remote, that Earth would be terminated and that existence would end. Lights out forever.

The primal know-the-enemy impulse took concrete form when the Johns Hopkins University Applied Physics Laboratory (APL) designed a spacecraft for NASA that was packed with instruments and that would be the first to be sent to an asteroid to study it in detail as part of the new Discovery program. As was mentioned in chapter 2, the asteroid candidate of choice was 433 Eros, a smooth, pockmarked near-Earth boulder that was thirty-three kilometers long and thirteen kilometers wide—about the size of Manhattan—and that was incongruously named after the Greek god of love (as in *erotic*).

The mission was at first called NEAR, for Near-Earth Asteroid Rendezvous, and then NEAR Shoemaker for the legendary godfather of planetary science. It entailed having the spacecraft land on Eros and stay on it—ride it—for a year, all the while collecting data on its prodigious bulk, composition, mineral content, internal mass, magnetic field, interaction with the solar wind and the space around it, and more. The spacecraft was launched from Cape Canaveral on February 17, 1996, flew by asteroid 253 Mathilde on June 27, 1997, and then went on to 433 Eros. After the first attempt at orbital insertion (as its handlers called it, which describes the adjustment of a spacecraft's momentum to enter into orbit around a space object) failed, it began flying ever-tighter circles around its quarry and started the first orbital study of an asteroid in February 2000. The visitor from Earth eventually landed on 433 Eros on Valentine's Day, February 12, 2001.

NEAR Shoemaker then began sending home everything it learned from its very close encounter. Not surprisingly, the asteroid's mineral composition was typical of other asteroids, and the tough guy had itself been the target of hits—"successive impacts"—in the shooting gallery. Andrew F. Cheng, who was also at the Johns Hopkins APL and who was a member of the science team, reported that there were many impact craters, most of them relatively large, and that the asteroid was "intact but deeply fractured," probably because it had broken off of an even larger rock.[13] The mission also showed that landing people on this or other large asteroids, either for scientific research or to mine them for rare minerals, was feasible. The spacecraft stopped sending data home on February 28, 2001, and an attempt to communicate with it in December 2002 was unsuccessful. It is still riding its asteroid, however, mute in the bitter cold.

The 433 Eros asteroid was no threat to Earth and never will be. But the mission provided the first detailed look at one of the large rocks in its element as representative of others, larger and smaller, that could be potentially hazardous or worse. The images of a scarred, potato-shaped boulder that was twice the size of Manhattan speeding through space were vivid reminders, yet again, that many dangerous objects prowl the neighborhood, far from home and near it. So the NEAR Shoemaker encounter, like the attack on Jupiter, remained lodged in the minds of the space-science community and those on Capitol Hill who were responsible for knowing about the potential danger posed by asteroids and comets that get too close.

The idea that asteroid and comet impacts are still a hazard had only just taken hold in 1980, as David Morrison has written, which is when the Alvarez team published what it found at the Chicxulub crater. Morrison adds that asteroids are statistically by far the greater threat than comets are; the threat from comets is comparatively miniscule.[14] But the threat from NEOs is potential, not immediate. The Wide-field Infrared Survey Explorer telescope, which was called "WISE" by its handlers at JPL, turned up 911 (that number again) of an estimated 981 NEOs larger than a kilometer, but none were thought to be potentially hazardous asteroids. That gestated in the scientific community and in Washington for a little more than a decade. And so, understandably, did the idea that what had come out of the sky to cause such death and destruction could come again, a point that is made to students in Astronomy 101. Those who are assigned *Contemporary Astronomy* by Jay M. Pasachoff, the standard text for that college-level course, learn that some Apollo asteroids—those that cross Earth's orbit—are potential impactors. "Most Apollo asteroids will probably collide with the Earth eventually," the author explains, "because their orbits may intersect the Earth's. Luckily there are only a few dozen Apollos greater than 1 km in diameter."[15] The widely used textbook—and justifiably so—might have pointed out that even Earth-crossers a lot smaller than a kilometer can, and have been, devastating.

With the Apollo Moon-landing program, NASA scored a monumental victory over the Soviet Union, followed by that enormously successful Solar System exploration program—the high point of which was unquestionably Voyager 2's twelve-year grand tour of four of the outer planets and the library of knowledge it sent home—and then the Cold War ended and a highly productive "space race" was over. That left NASA with a conundrum. It became the victim of its own success. All of those accomplishments were by definition spectacularly unique, but repeating them was considered irrelevant and wasteful. The resources were there, but there was no mission. So NASA became a powerful and famously inventive agency without a major long-term goal that was inherently dramatic enough to hold the public's (and congressional appropriations committees') interest. There is such a mission, however. It is planetary defense done in conjunction with the international community.

The Alvarez group's discovery at Chicxulub was widely accepted as certain evidence of Earth's vulnerability to bombardment from space: that an object larger than the one that caused the long global winter that made the dinosaurs extinct and killed off so many other species could bring on Doomsday, making virtually all life on the planet extinct. It should be noted that the extinction theory was widely, but not universally, accepted. It was challenged by an international group of seven scientists with excellent credentials, led by Gerta Keller, a geoscientist at Princeton University, who maintained that the impact at Chicxulub happened about three hundred thousand years before the K-T boundary, as the dinosaur-ending Cretaceous-Tertiary time division is called in shorthand, and that it was therefore not responsible for their extinction.[16]

The theory that the giant reptiles and a lot of other living things were done in by the NEO that simultaneously caused the great boundary still held among the science community at large and the public, though. The K-T event killed every creature larger than a sheep that did not live near a river or in shallow water or that was not deep in a cave or underground. But many beasts survived the impact, including small dinosaurs that literally took flight and evolved into birds. Having a vague knowledge of the mass extinction and understanding that the space agency had the technological wherewithal—at least the basics of it—to address the distinct possibility that it could happen again, or even be worse, the House of Representatives stipulated the following in its NASA Multiyear Authorization Act of 1990:

> It is imperative that the detection rate of Earth-orbit-crossing asteroids must be increased substantially, and that the means to destroy or alter the orbits of asteroids when they threaten collision should be defined and agreed upon internationally. The chances of the Earth being struck by a large asteroid are extremely small, but since the consequences of such a collision are extremely large, the Committee believes it is only prudent to assess the nature of the threat and prepare to deal with it. We have the technology to detect such asteroids and to prevent their collision with the Earth.[17]

The committee therefore ordered NASA to conduct two workshops: one to dramatically increase the detection rate of Earth-crossers, and the other to define the systems and technologies that would be required to alter their courses or destroy them if they posed a danger to life on Earth. Since the danger was understood to be global, the committee prudently recommended international participation.[18]

The first workshop had twenty-four men and women from six countries—the United States, Russia, France, Finland, India, and Australia—who amounted to the NEO first team. Individuals included David Morrison, Clark Chapman, Donald Yeomans, Gene Shoemaker, Tom Gehrels of the University of Arizona, Brian Marsden of the Harvard-Smithsonian Center for Astrophysics in Cambridge (which had long kept track of NEOs), Louis Friedman of the Planetary Society, and Duncan Steel, an Australian astronomer whose *Rogue Asteroids and Doomsday Comets* was an easily understood book written for a lay audience. They, together with Alexander Basilevsky, the director of the Laboratory for Comparative Planetology at the

Russian Academy of Sciences' Vernadsky Institute in Moscow, who was not at the meeting, were a cadre of the world's leading experts on the objects, large and small, that swarm around it. The idea was to inform as many people as possible about the remote but real danger that lurks out there.

The group produced an encyclopedic report in 1992 called the Spaceguard Survey that described the situation, including the potential troublemakers in substantial detail, and was a crucial first step in the defense of the planet. As everyone in the community knew, logic required that the NEOs needed to be defined and described before the threat they posed could be calculated and articulated so an effective defense could be mounted. "There are two broad categories of objects with orbits that bring them close to the Earth: comets and asteroids. Asteroids and comets are distinguished by astronomers on the basis of their telescopic appearance," the introduction explained. "If the object is star-like in appearance, it is called an asteroid. If it has a visible atmosphere or tail, it is a comet. This distinction reflects in part a difference in composition: asteroids are generally rocky or metallic objects without atmospheres, whereas comets are composed in part of volatiles (like water ice) that evaporate when heated to produce a tenuous and transient atmosphere [or "tail"]. . . . For our purposes, the distinction between a comet and an asteroid is not very important. What matters is whether the object's orbit brings it close to the Earth—close enough for a potential collision."[19] Earth-crossing asteroids (ECAs)are those that cross Earth's path as it circles the Sun the way torpedoes pass in front of a ship—they are the ones that are most likely to collide with Earth, and they are therefore the most closely watched.

A chapter of the survey was devoted to the hazard of impacts, and specifically to the relationship between the size of the impactor and the damage it would likely cause, from local destruction to the threshold size for a global catastrophe.

Asteroids of stony or metallic composition that are larger than 100–150 meters can make it to the ground intact and blast out relatively small craters, while those that are larger than 150 meters can make craters that are three kilometers in diameter. Worse, the larger-than-150-meter asteroids' zone of destruction would extend well beyond the impact area, damaging or flattening buildings with the shock wave from their explosions. The good news is that they come along only about once every five thousand years. And the planet is very likely out of harm's way for at least another five generations. "No object that is now known has an orbit that will lead to a collision with our planet during the next century, and the vast majority of the newly discovered asteroids and comets will also be found to pose no near-term danger. Even if an ECA has an orbit that might lead it to an impact, it will typically make hundreds of moderately near passes before there is any danger, providing ample time for response. However, the lead time will be much less for a new comet approaching the Earth on a long-period orbit," the report explains. The authors claim that no object that is now known, that is being tracked, is in an orbit that will lead to a collision with Earth during the next century. That would undoubtedly draw sneers from the residents of Chelyabinsk and probably an expression of astonishment from Michelle Knapp.

And it would very likely dumbfound the people of the Eastern Mediterranean. What happened there on June 6, 2002, was a real attention-getter. There was an explosion by an undetected asteroid between Libya and Crete that had the power of a small atomic bomb. It caused no damage and, since it occurred over the sea, no fragments were recovered for study (or sale). But Gen. Simon P. Worden, the deputy director of operations at US Space Command (which scrupulously tracks manmade objects such as spacecraft circling Earth, as well as some objects that are made elsewhere), noted that had the explosion occurred over or near the Indian subcontinent a few hours earlier it could have started a nuclear war between India and Pakistan, both of which were having a military standoff with tensions running even higher than usual. The blast might have been interpreted by either side as the prelude to an attack by the other, and either could have answered with an immediate, reflexive counterattack.

"A few weeks ago the world almost saw a nuclear war," Worden said in a speech on July 10. He continued,

> Pakistan and India were at full alert and poised for a large-scale war—which both sides appeared willing to escalate into nuclear war. The situation was defused—for now! Most of the world knew about this situation and watched and worried. But few know of an event over the Mediterranean in early June of this year that could have had a serious bearing on the outcome. U.S. early warning satellites detected a flash that indicated an energy release comparable to the Hiroshima burst. We see about 30 such bursts per year, but this one was one of the largest we've ever seen. The event was caused by the impact of a small asteroid—probably about 5–10 meters in diameter on the earth's atmosphere. . . . The event of this June caused little or no notice as far as we can tell. But had it occurred at the same latitude, but a few hours earlier, the result on human affairs might have been much worse. Imagine that the bright flash accompanied by a damaging shock wave had occurred over Delhi, India or Islamabad, Pakistan. . . . The resulting panic in the nuclear-armed and hair-trigger militaries there could have been the spark that would have ignited the nuclear horror we've avoided for over a half century. This situation alone should be sufficient to get the world to take notice of the threat of asteroid impact.[20]

Worden was and remains mindful of the NEO threat and is therefore a strong proponent of robotic and crewed missions to them to increase our knowledge.

By then, the United Nations had long since been concerned about the threat. It held a near-Earth-objects conference at UN headquarters and at the Explorers Club in New York in April 1995, at which many of the astronomers, earth and planetary scientists, and those specializing in astronautics who had begun to devote their professional lives to studying the problem made presentations that fundamentally helped to shape the international planetary-defense program. They included Yeomans, Marsden, Ostro, Chapman, Keller, Morrison, and Gehrels. The conference also included John L. Remo of QuantaMetrics, a company that specializes in selling and marketing to scientific researchers; Mark B. E. Boslough, a physicist at the Sandia National Laboratories and an expert on planetary impacts who had made many science documentaries and television appearances; Michael R. Rampino and Bruce M. Haggerty, professors in New York University's Earth and Environmental Science Program; and several other scientists from around the world. In all, they gave forty-seven presentations during the three-day session that, as custom required and would continue to require, first described the NEO "population" and then focused on ways to detect and then mitigate them.[21]

Given that there are far fewer really menacing asteroids and comets out there because of the relative rarity of the very big ones, there is indeed breathing time, provided something terrible does not come out of the Oort Cloud unexpectedly. The cloud is believed, in theory, to envelop this Solar System—it has not been seen as more substantial parts of the universe have—and contain billions of comets. A kilometer-size impactor would cause destruction in the megaton range (a term that inevitably suggests a comparison with the effects of nuclear weapons).

Larger attackers—those measuring one to five kilometers—would cause serious global consequences, the Spaceguard Survey reported, and specifically gouge out craters up to fifteen times the diameter of the projectile. The primary hazard that would come from such a strike—provided no people or property were hit—would be a global "veil of dust" in the stratosphere that could cause massive, worldwide crop failures that could threaten civilization. That added a dimension to the threat. Damage had always been characterized as immediate death and destruction: cities or regions destroyed and mass fatalities. But now a more subtle element of the equation was added: starvation. The authors admitted that the threshold size of the impactor that would cause such a "global catastrophe"—its minimum size—is not accurately known. And contrary to the popular portrayal of asteroids as smooth, potato-shaped rocks like 433 Eros, they are most often jagged since they are chunks of rocks or metal—fragments—that were violently broken off of larger bodies, including planetoids, which are very large, pretentious asteroids.

As every police officer, sports coach, physician, lawyer, and businessman or businesswoman knows, competition requires knowing the opponent, and the more thoroughly, the better. That certainly applies to planetary defense—to competition with NEOs—which is why the Spaceguard Survey called for increased observation, notably by optical astronomy, radar astronomy, and physical observation with photometric equipment. In common with other asteroid and comet watchers around the world, its collective membership also called for international cooperation. "That the hazard posed by NEO's is a problem for all humankind hardly needs repeating," they repeated. "The likelihood of a particular spot being the target of an impact is independent of its geographic position, so that we are all equally at risk."[22]

The community has been well aware of the danger for decades. Congress got on board in 1998 after hearings were held that May. There, Clark Chapman, William Ailor of the Aerospace Corporation, Gregory Canavan of the Los Alamos National Laboratory, John Lewis of the University of Arizona, and Carl Pilcher, the director of NASA's science program, testified about the threat and what was being done to address it.

Pilcher told the Subcommittee on Space and Aeronautics that NASA was so concerned about what its and other telescopes picked up that it was committed to cataloging within a decade 90 percent of all NEOs with a diameter larger than one kilometer and that the program was on track to do so. He added that the budget for completing this task had been doubled to $3 million and that that allocation—relatively meager as it was—would at a minimum be maintained. Canavan told the subcommittee that improved technology had increased the detection rate, but that long-period comets—those that cross Earth's path—remained a serious concern and could constitute as much as half of the NEO threat. He also repeated the axiom that asteroids had to be characterized (meaning adequately defined) for their threat to be reduced, and he called for cooperation between the Department of Defense and NASA, as happened in the Clementine II mission in 1994, when they combined to test spacecraft components and sensors by sending the spacecraft to observe the Moon and asteroid 1620 Geographos. Usable data came back from the Moon but not from the asteroid because of a technical malfunction. Ailor discussed the Leonid meteor shower and the danger it would pose to satellites when it started in November. It was a spectacular show and caused no apparent harm to the Earth-circlers. And Lewis sounded a distinctly upbeat note by telling the subcommittee that asteroids were economically valuable because they could be mined, and he carefully pointed out that the keys to successful exploitation for minerals were lower launch costs and carefully choosing asteroids that are the most accessible and have the richest mineral concentrations.[23]

The congressmen, clearly impressed by the presentations, reacted by reaffirming what NASA was already doing. They formally mandated it to catalog within a decade—by 2008—90 percent of near-Earth asteroids (NEAs) a kilometer or larger, and that became the Spaceguard Survey. The name memorializes Arthur C. Clarke: Saint Arthur, the Visionary. Many others have decided to pay homage to Clarke the same way. The International Astronomical Union's Working Group on Near-

Earth Objects held a workshop in 1995, beginning the Space Survey, which led to the creation of the private, international Spaceguard Foundation that was started in Italy in 1996 and that is dedicated to discovering and studying NEOs. There is also a Japan Spaceguard Association (which is based in Tokyo) and the Bisei Spaceguard Center (also in Japan), which do the same thing. And Spaceguard UK operates the Spaceguard Centre in Knighton, which was set up to provide information on asteroid and comet impacts, to find ways of predicting them, and to get that information to the news media and educational institutions.

The relationship between Congress and NASA on identifying large NEOs became somewhat incestuous. The space agency announced that it planned to find and identify 90 percent of all NEOs that potentially threatened Earth. Then, as noted, Congress dutifully reacted by ordering NASA to find and identify 90 percent of all Earth-crossers that endangered the planet. The space agency reacted to the congressional order by proclaiming in September 2011 that, lo and behold, it had met the congressional goal and had found more than 90 percent of the rocks that could cause a planet-wide catastrophe.[24]

Not to be left out, the US Air Force issued a report, *Preparing for Planetary Defense: Detection and Interception of Asteroids on Collision Course with Earth*, a thirty-three-page canon that recommended that the Department of Defense take an active role in planetary defense. It credited Congress with understanding the danger and ordering the creation of the International NEO Detection Workshop that produced the Spaceguard Survey and an NEO Interception Workshop that met in 1992 to come up with ways to intercept, deflect, or destroy anything that appears to be on a collision course with Earth. Putting the situation in perspective, the report opened by matter-of-factly asserting the following:

Most humanity is oblivious to the prospect of cosmic collisions, but this hazard from space is a subject of deadly concern to the population of the planet. Work by several nationally recognized scientists who have been investigating this issue for a number of years, some for decades, has brought an awareness that, to the average citizen of the U.S., the risk of death may be just as great from an asteroid strike as from an aircraft accident. Those unfamiliar with these studies may find this incredulous when, in fact, there have been no recorded deaths due to asteroid strikes, albeit there have been close calls from small meteorites striking cars and houses. However, the probability is finite, and when it occurs, the resulting disaster is expected to be devastatingly catastrophic. But because we are dealing with events, time scales and forces well beyond the human experience, the threat is not universally recognized.[25]

With the public's ignorance established, the report went on to describe the danger, citing the Alvarez group's work, the extinctions, and comet Shoemaker-Levy 9's impending attack on Jupiter. Then it, too, called for an enhanced capability to find and characterize potential threats and come up with ways to prevent collisions. There are two ways to do that, the report stated: (1) propulsion, meaning a frontal attack to stop the asteroid by sending a rocket into it or by using nuclear energy and ultimately some hypervelocity, or antimatter weapons; and (2) deflection, which would involve nuclear and kinetic energy, lasers, and even putting solar sails on them that would gently move them off course.[26]

The august Federation of American Scientists put out its own report, which had the same title as the Air Force report and which essentially said the same thing, as well as what was in the NASA Spaceguard Survey: the threat had to be clearly defined and then mitigated with an international response.

The American Institute of Aeronautics and Astronautics (AIAA), which was started in 1963 as an amalgamation of the venerable and widely respected American Rocket Society and the American Interplanetary Society, has some thirty-five thousand individual members and ninety corporate members. It therefore wields considerable power in the aerospace world and weighed in with a position paper in October 2004 whose title, "Protecting Earth from Asteroids and Comets," got right to the point. It led off by suggesting that an organization be created within the US government that would be specifically responsible for planetary defense and would be an interagency office charged with dealing with "all aspects of Planetary Defense." It also proposed that a senior-level inter-agency working group be formed to define the appropriate makeup and reporting structure of the organization, develop a plan that would lead to its creation, and procure funding for its support. The AIAA paper also called for the office to establish a formal procedure for getting word out when the probability of an impact exceeds specified thresholds. And since the threat is global, it recommended starting a dialogue among other nations and international institutions to characterize the challenges that would be involved in an international deflection program. Furthermore, it called for broadening the Spaceguard Survey to include one-hundred-meter and larger NEOs; get more information about asteroids (including by missions to them so deflection techniques could be developed); conduct actual flight tests to demonstrate the ability to change a potential impactor's orbit; and—getting to the human factor—sponsoring research to assess the political, social, legal, and disaster-relief consequences of a serious NEO threat, mitigation effort, or possible impact.

"While noteworthy efforts are being made to detect threatening objects," the paper concluded, "Earth is effectively blind to NEO objects of a size range that could lead to immediate and long term deaths of thousands to millions of people and is unprepared should a short term threat be detected."[27]

By suggesting that a national organization be created that would be responsible for overseeing the asteroid and comet threat, the institute was clearly implying that NASA had too

many other projects and programs to be able to adequately concentrate on this most important mission. That was not the case,
though. If anything, the space agency needed a major program
that would fill the void left by the ending of Apollo in 1972. It
therefore decided that planetary defense should be handled by
the space agency. Congress agreed. By 2005, the legislators were
so concerned about the asteroid and comet population that the
NASA authorization act substantially lowered the minimum
size that had to be located and studied to 140 meters. Its charge
to NASA clearly reflected its concern:

> The Congress declares that the general welfare and security of the
> United States require that the unique competence of the National
> Aeronautics and Space Administration be directed to detecting,
> tracking, cataloguing, and characterizing Near-Earth Asteroids and
> comets in order to provide warning and mitigation of the potential
> hazard of such Near-Earth Objects to the Earth.
>
> The Administrator shall plan, develop and implement a Near-
> Earth Object Survey program to detect, track, catalogue and char-
> acterize the physical characteristics of Near-Earth Objects equal to
> or greater than 140 meters in diameter in order to assess the threat
> of such Near-Earth Objects to the Earth. It shall be the goal of the
> Survey program to achieve 90 percent completion of its Near-Earth
> Object catalogue (based on statistically predicted populations of
> Near-Earth Objects) within 15 years after the date of enactment of
> this Act.[28]

That officially started the Spaceguard Survey, which stands as a
milestone in the new quest for planetary defense.

The National Research Council (NRC), which is the investigative division of the National Academy of Sciences, the
nation's preeminent and ultra-prestigious science body, also
took an interest in the menacing meteors and their icy cohorts.
Its fourteen-member Survey and Detection Panel held three-day
fact-finding hearings in Washington, Tucson, and Santa Fe in
2009, at which scientists, other specialists, and a NASA representative made presentations that described the NEO situ-

ation in copious detail. As was customary, the presentations described what comets, asteroids, and their low-flying derivatives, meteors, are; what threat they pose; and how that threat can be (yes . . .) mitigated. Tucson may have been picked so attendees could supplement the scholarly presentations with a day trip to nearby Meteor Crater to see firsthand what the impactors their colleagues were describing could do.

Maj. Lindley N. Johnson, the NEO Observations Program executive of NASA's Planetary Science Division, presented an overview of the NEO situation. His presentation included an outline of the NEOs' history and what effects they would have, based on their size and average impact interval, starting with a puny meteor under fifty meters that would break up high in the atmosphere and cause no damage and progressing to ever-larger ones. Larger space objects that are bigger than fifty meters would cause a Tunguska-like event; bigger than 140 meters would cause a regional event; a kilometer or bigger would have a relatively serious global effect; and one that is ten kilometers or bigger, which occurs only once every one hundred million years would pack one hundred million megatons of explosive power and turn off the lights. That is, it would cause an extinction event. And Johnson mentioned a near-Earth asteroid known as 2004 MN4 Apophis (later renamed 99942 Apophis), which, as was discussed earlier, is appropriately named after the Egyptian serpent god that is the lord of chaos and darkness. It flies past Earth every seven years and, in 2013, came within nine million miles. The Apophis meteor caused some anxiety in 2004 when it was thought that there was a very slight possibility, soon shown to be wrong, that during its expected approach, it would crash into Earth in 2029. It will come around again in 2036, though astronomers have concluded that it will not be a threat then, either. It is therefore not a dreaded PHA: potentially hazardous asteroid.

In 2007, the Planetary Society organized a $50,000 competition to design an unmanned probe that would find Apophis and follow it for almost a year, collecting data that could help determine whether it posed enough of a threat to warrant a mission that would deflect it. The society received thirty-seven entries from twenty countries and picked a design called "Foresight" by SpaceWorks Enterprises, a large US firm, which proposed a simple spacecraft that would orbit the asteroid for a month and then follow it around the Sun for ten more months to get enough information about the meteor's trajectory so the likelihood of an eventual impact could be calculated.[29]

In January 2013, four years after the meeting in Santa Fe, NASA decided that Apophis will make a relatively close approach in 2029—19,400 million miles—and will swing by again in 2036 at a greater distance, meaning a collision was ruled out. Since the asteroid is 325 meters in diameter, that was very good news indeed, because an impact would cause an explosion in the five-hundred-megaton range.[30]

The research presented at the three NRC meetings was duly published in 2010 in a definitive, understandable, and well-illustrated 144-page report titled *Defending Planet Earth: Near-Earth-Object Surveys and Hazard Mitigation Strategies* that had sections on risk analysis, the survey and detection of NEOs, their characterization by various observation techniques, ways to mitigate the threat, the need (again) for national and international cooperation, and what is required in the way of financial investment. (Ten million dollars a year would allow the current program to continue while, at the other end of the investment spectrum, $250 million annually would buy a robust program with redundant systems that would combine ground- and space-based observation and research on impact techniques for changing the orbit of a threatening asteroid: impacting an impactor, so to speak.)

"Impacts on Earth by Near-Earth Objects (NEOs) are inevitable," the report stated at the top of the chapter on mitigation. "The impactors range from harmless fireballs, which are very frequent, through the largest airbursts, which do not cause significant destruction on the ground, on average occurring once in a human lifetime; to globally catastrophic events, which are very unlikely to occur in any given human lifetime, but are probably randomly distributed in time." The report described the usual mitigation possibilities: slowly pushing or pulling the attacker off course, a kinetic impact that would shove it in another direction, or a nuclear blast that would decisively change its orbit. And in the event that any or all of those mitigation strategies failed, the report advised that a civil defense system be set up to evacuate a region that takes a small but direct hit.[31]

What Robert F. Arentz of the Boulder, Colorado–based Ball Aerospace and Technologies Corp., who was on the National Research Council's Survey and Detection Panel warned bears repeating: "It's not a matter of if," he said, "it's a matter of when."[32]

4

THE FASCINATION FACTOR

The prospect of Doomsday, or at least the threat of it, has fascinated people from time immemorial for three fundamental reasons.

For one thing, preventing it brings out the best of the human species, at least in terms of resourcefulness. Triumph over a formidable opponent is deeply pleasing because it honors and dignifies the human spirit. That is why underdogs who win are celebrated while there are usually no celebrations when overdogs (if they can be called that) are victorious. It is a formula that has always been understood by the coaches of teams that play competitive sports, by writers who craft both fiction and nonfiction stories, and by the producers and writers of war films who have the heroes suffer physical and sometimes mental injuries before they finally defeat the enemy by dint of sheer courage and reserve strength.

Then, too, the end of humanity would deprive the universe of wonderful and probably unique creatures (with all their apparent flaws). That makes us feel special and makes our fighting for survival rather than succumbing to extinction of the utmost importance. "Perhaps a sense of impending doom is necessary in certain situations to overcome complacency that might come with a misunderstanding of potential dangers," Michael Moyer, an editor at *Scientific American*, has suggested. "If people all over the world had not been overwhelmed by the concern of possible extinction during the Cold War, there may have been a nuclear war with dire consequences. Even now this concern has to be constantly reinforced."[1]

Most religions, on the other hand, assure their followers that belief in a supreme being will bring salvation—as it did to Noah and the creatures on his ark—and that the faithful who suffer will have eternal life, as did Jesus Christ. Death, even when it comes after excruciating pain, is therefore merely physical, not spiritual, and so it can be endured because it ends in immortality.

Both belief systems—the triumph over death and succumbing to it for spiritual salvation—make the possibility of Doomsday (that is, the end of the world and of all the life on it because of catastrophic destruction) endlessly fascinating. Whether the end can be prevented by bravery and intellect or accepted as the way to gain entry to heaven, it has always captivated a very large segment of humanity and continues to do so. Those people, in the many millions, are an eager audience for writers and others who describe the abiding dangers that threaten our existence and have the characters either fall victim to them but somehow survive anyway or find ways to thwart them, defeat them, by being imaginative, resilient, and heroic.

Homer (if it was Homer) was among the first, at least in the West, to glorify those traits in the *Iliad* and then in the *Odyssey*, in which Odysseus is triumphant over his powerful and implacable enemy, Polyphemus the Cyclops (the one-eyed son of Poseidon, god of the sea), because of his bravery and tenacity. Odysseus speaks of the transcendent virtue of literally as well as figuratively staying the course (or hanging in, as we have come to call it):

> Yea, and if some god shall wreck me in the wine-dark deep, Even so I will endure. . . . For already have I suffered full much, And much have I toiled in perils of waves and war. Let this be added to the tale of those.[2]

Given time (and returning home to domestic tranquility), the veteran of the war with the Trojans might have organized

a coalition among all the city-states that would have protected
them from the Cyclops and ultimately killed it with a system
that could have been called SEAGUARD. Homer would have
appreciated what Arthur C. Clarke wrote in a distinctly similar
vein, so it bears repeating:

> Six hundred thousand people died, and the total damage was more
> than a trillion dollars. But the loss to art, to history, to science—to
> the whole human race, for the rest of time—was beyond all compu-
> tation. It was as if a great war had been fought and lost in a single
> morning. . . . After the initial shock, mankind reacted with a deter-
> mination and a unity that no earlier age could have shown. Such
> a disaster, it was realized, might not occur again for a thousand
> year—but it might occur tomorrow. And the next time, the con-
> sequences could be even worse. Very well: *there would be no next
> time* . . . So began Project SPACEGUARD.[3]

That is ennobling. It is what makes Darwin not just scientifically important but, on a subtle level, inspiring. Survival of
the fittest creatures is accomplished not by chance but through
superiority; by a determination to overcome mortal challenges
and either coexist within the environment or, if sheer existence
is threatened, to dominate, vanquish, or destroy the challenger.
The long history of warfare offers innumerable examples of the
fate of those who lacked the will or capability to assure their
survival and freedom. They were at the mercy of the enemy
when war came, and the price they paid was subjugation,
ruination, and often death.

Clarke and other science fiction writers have transgressed
war between nations—mere political entities; the new city-
states—and shown that the whole planet is at war with nature,
with its environment, and has been since its creation. In that
grand scheme of things, military conflicts between "sovereign"
states are trivial since, where nature is concerned, they are in
fact not sovereign at all. Ultimately, they exist at the whim of
the outer world, of the universe, that surrounds them. They are

simultaneously nurtured by nature and at war with it, and the weapons that are arrayed against them extend from infectious microbes to some creatures on land and sea, to the weather, to the bowels of the planet itself, and to the large projectiles from space that constitute the hail of bullets. The large rocks and comets are the most interesting because, unlike diseases, they come from somewhere else—other worlds—so they are exotic, and there is no possibility of inoculation (at least not yet, if that metaphor is valid). And they are by far the most dangerous, even when all-out nuclear war is taken into account, since only one of them could end everything.

Clarke, who was highly knowledgeable about science and the environment in which Earth exists, understood that. So did Isaac Asimov and Robert A. Heinlein who, with Clarke, were the scientifically informed triumvirate of what is now called the Golden Age of Science Fiction. Clarke's *The Exploration of Space*, *The Exploration of the Moon*, *The Making of a Moon*, and *The Promise of Space*, among many other nonfiction works, initiated untold millions into the space fold and its infinite dimensions.

So did a science fiction masterpiece called *2001: A Space Odyssey*, a novel that grew out of a short story called "The Sentinel." Film producer Stanley Kubrick picked up the novel and, along with Clarke, turned it into a film of the same name. The story is about a mission to Jupiter in which there are encounters with mysterious black monoliths that seem to be affecting human evolution. But there is a dramatic subplot about one of man's robotic creations, ostensibly his computerized mechanical servant, HAL 9000, that rebels and tries to assert its superiority by attempting to kill two astronauts while they are outside the spaceship on a mission beyond Jupiter. It succeeds in killing one of them, but the other makes it back inside with the other astronaut's corpse and methodically disconnects HAL's wiring while it gently pleads for its life to no avail. It was an effective dramatization of the old theme of humankind's robotic creations rebelling, thinking for themselves, and turning on their masters, which was started by Mary Shelley when she published *Frankenstein; or, the Modern Prometheus* in 1818. The film version of *2001: A Space Odyssey*, whose screenplay was written by Kubrick and Clarke, had mixed reviews but developed a cult following when it came out in 1968. Perhaps this was partly because, by sheer coincidence, it was followed by three astronauts on the Apollo 8 mission who became the first humans to see the far side of the Moon and an earthrise beyond the lunar horizon. They sent home television pictures of it for all the world to experience. The short story "The Sentinel," the full-length novel derived from it, and the subsequent classic film stand as examples that show what imaginative and resourceful humans can do to survive their own technology and an intensely hostile environment.

That environment was described in harrowing detail by Larry Niven and Jerry Pournelle in *Lucifer's Hammer*, which was published in 1977 and, thus, among the first works of fiction to try to anticipate how the survivors on a planet that is nearly annihilated handle the cataclysm and adapt to what it causes: a new ice age. In *Lucifer's Hammer*, a comet is spotted by a wealthy amateur astronomer named Tim Hamner and turned into a potential media event when it is determined that the boiling ice will pass close to Earth. After its discovery, a California senator named Arthur Jellison gets an Apollo-Soyuz mission to study "The Hammer," as it is christened by the news media.

But then the comet breaks up, dumbfounding professional astronomers at NASA's Jet Propulsion Laboratory (JPL) who were unable to track it, and it hits Earth like buckshot. Large chunks strike parts of Europe, Africa, the Gulf of Mexico, and the Atlantic and Pacific Oceans. Those that hit land set off volcanoes and start earthquakes around the world, including one along the San Andreas Fault that severely damages California.

To make matters worse, the ones that hit the oceans cause tsunamis, inundating coastal cities, including Los Angeles, and millions perish. Hundreds of millions of hapless men, women, and children are exterminated around the world like vermin. Plagues break out and the climate changes, causing weeks of rain, which leads to flooding that wipes out crops and that, in turn, forces otherwise decent folks to scavenge, steal, use weapons for self-protection, eat rats and human corpses, and ultimately resort to cannibalism. Fearing that the new ice age is going to send desperate Russians south for warmth and food, China launches a preemptive nuclear attack on Russians cities, but the Caucasians stick together in the face of another Yellow Peril. Russia and the United States retaliate, effectively destroying China.

Meanwhile, Senator Jellison and other landowners create fiefdoms within his "Stronghold" in which ordinary workers are forced into subsistence farming like serfs while an evangelist named Henry Armitage teaches that The Hammer's arrival signals the joyful End Times. That is a theme that will recur in both apocalyptic fiction and fact. Jellison takes advantage of the crisis to become the Stronghold's strongman. He seizes control of the remnants of the US Army and, with bikers and other gang members, starts the New Brotherhood Army, which is headquartered in the Stronghold and maintains order through military discipline. Jellison has made himself the lord and master of his world. In that situation, the status of the police is diminished, and since he now makes the law, lawyers are unnecessary. But Hamner and his wife, Eileen, are determined to endure the carnage. And they do, showing that the survival instinct can triumph over even a cosmic catastrophe like being hit with the devil's gigantic hammer.

Reviewing *Lucifer's Hammer* for *Library Journal*, Judith T. Yamamoto said that it was full of "good, solid science, a gigantic but well developed and coordinated cast of characters, and about a megaton of suspenseful excitement." She remarked that the pro-technology pitch might turn off some readers, but "all in all it's a good book, if not a great one."[4]

Many readers agreed. "This is 5-star sci-fi all the way! If all you read is the first 100 pages, however, you probably won't agree with that. You see, the first part of the book is a bit slow in getting moving, but that's because the authors introduce a whole string of characters [who] interact with one another as the story unfolds. And once the action starts, it doesn't stop. In fact, it makes you want to store some food, some water, some other things . . . and get ready for what COULD happen," one wrote to Amazon.

> As I started reading this book, I thought to myself, this book has many similarities with the movie *Deep Impact*. Was I ever wrong with that assumption! This book goes way beyond *Deep Impact*. It goes beyond it in that this book is not so much about events surrounding a comet-earth collision as it is about the aftermath, and how people do or do not cope with that kind of calamity. Imagine this . . . world-wide cataclysmic events wipe out the major governments on the planet—national, state, and local governments collapse, and people are left to fend for themselves. What will they do for food, shelter, personal safety, information, etc.? It's a whole new ballgame out there! The kinds of challenges described in the book bring out the best in some people, the worst in others, and trapped in the middle of everything that's happening are the characters you'll come to know quite well.[5]

"The gigantic comet had slammed into Earth, forging earthquakes a thousand times too powerful to measure on the Richter scale, tidal waves thousands of feet high," a reviewer in the *Cleveland Plain-Dealer*, who seems to have savored the calamity, reported appreciatively. "Cities were turned into oceans; oceans turned into steam. It was the beginning of a new Ice Age and the end of civilization. But for the terrified men and women chance had saved, it was also the dawn of a new struggle for survival—a struggle more dangerous and challenging than any they had ever known."[6]

Clarke brought his own hammer to science fiction, though he chose to protect Earth from terrible destruction by pushing the approaching devastator off course rather than forcing the planet to endure it and survive, as Niven and Pournelle had (he had read *Lucifer's Hammer*). That, of course, is precisely what NASA, the US Air Force, the world's other space agencies, and the international space community, very much including the B612 Foundation, want to do.

And he did it in *The Hammer of God*, which was published in 1993, nine years before the B612 Foundation came to be, and was only the second work of fiction to be published in *Time* magazine. Like Niven and Pournelle, he used the hammer analogy in that book and in *Rendezvous with Rama*, the latter of which is about a huge alien spacecraft heading toward Earth that is at first mistaken for very big rock. Clarke opened with what is effectively a brief prologue about what happened at Tunguska and then turned it into fiction. "Moving at fifty kilometers a second, a thousand tons of rock and metal impacted on the plains of northern Italy, destroying in a few flaming moments the labor of centuries. The cities of Padua and Verona were wiped from the face of the Earth; and the last glories of Venice sank forever beneath the sea as the waters of the Adriatic came thundering landward after the hammer blow from space."[7] Hammers are for pounding things, after all, from nails to planets.

The seed that sprouted into *The Hammer of God* seems to have been planted by Zanoguera and Penfield's discovery of the crater at Chicxulub, followed by the Alvarezes' work there. Clarke knew father and son and wrote this "puff" (as he put it) for the jacket of Luis Alvarez's autobiography:

> And now he's engaged on his most spectacular piece of scientific detection, as he unravels the biggest whodunit of all time—the extinction of the dinosaurs. He and his son Walter are sure they've found the murder weapon in the Crime of the Eons. . . .

Since Luis's death, the evidence for at least one major meteor (or small asteroid) impact has accumulated, and several possible sites have been identified—the current favorite being a buried crater, 180 kilometers across, at Chicxulub, on the Yucatan Peninsula.

Some geologists are still fighting stubbornly for a purely terrestrial explanation of the dinosaur extinction (e.g., volcanoes), and it may well turn out that there is truth in both hypotheses. But the Meteor Mafia appears to be winning, if only because its scenario is much the most dramatic.[8]

With Clarke's words in mind, it's not hard to see the connections between the Alvarezes' work and *The Hammer of God*. The novel takes place in 2110, when a spaceship called *Goliath* is sent out to meet an asteroid named Kali that is discovered by an amateur astronomer on Mars to be heading in the direction of Earth. The astronomer, Dr. Angus Millar, is bored because there are no exotic diseases on Mars as there are on Earth. (In fiction, amateur astronomers apparently discover more threatening objects in space than professionals do—"When Worlds Collide" being a notable exception—perhaps because that adds an element to the story with which average readers, who may be intimidated by professional astronomers, can readily identify.) Recalling how excited he had been as a boy in 2061, when he saw Halley's Comet return, he builds an instrumented telescope and notices the approaching big rock, a potential Earth-annihilator:

It was an asteroid, just beyond the orbit of Jupiter. Dr. Millar set the computer to calculate its approximate orbit, and was surprised to find that Myrna—as he decided to call it—came quite close to Earth. That made it slightly more interesting.

He was never able to get the name recognized. Before the IAU could approve it, additional observations had given a much more accurate orbit.

And then only one name was possible: Kali, the goddess of destruction.

Kali is the name of the Hindu goddess of empowerment, which is to say, of life and death. And The IAU is the International Astronomical Union, a real organization headquartered in Paris, France.

Clarke was certain to add to his credibility by adroitly basing his fiction on certifiable fact, which would impress both those who knew the science and those who did not but appreciated it, however abstractly. This simple but powerful message made the point up front:

> All the events set in the past happened at the times and places stated: all those set in the future are possible. And one is certain. Sooner or later, we will meet Kali.

Indeed. There is absolute agreement among professional astronomers and virtually everyone in the international space community that, asteroid and comet traffic in the neighborhood being what it is, it is not a matter of if; it is a matter of when.

And, like Niven and Pournelle, Clarke effectively used religion as an evil counterpoint to the attempt to save Earth from the Apocalypse. The believers in God are diabolical fanatics who call themselves "The Reborn" and who try to sabotage the mission because they want Kali to destroy Earth. They want the Apocalypse because they will be able to shed their bodies, their physical existence, so their Lord will grant their spirits eternal life in heaven. They are therefore convinced that trying to head off Kali and save the planet is not good but blatantly sinful. For those who want to be reborn, Kali is God's messenger and their savior, so trying to prevent the collision is considered sacrilegious and reprehensible. As usual, Clarke did his homework. This is from 2 Peter 3:10–13 in the King James Bible: "But the day of the Lord will come as a thief in the night; in which the heavens shall pass away with a great noise, and the elements shall melt with fervent heat, the earth also and the works that are therein shall be burned up."

As all fiction writers know, the old hackneyed expression still applies: there can be no good without evil and no gripping story without conflict. Imagine US marshal Will Kane turning in his badge and riding that buggy out of town for a honeymoon with his Quaker pacifist bride in *High Noon*, with Frank Miller and his brothers still heading for town to gun Kane down. The beleaguered lawman is in a very tense situation, since he is substantially outnumbered and outgunned and must fight his enemy or risk being killed.

So must Earthlings as *Goliath*'s skipper, Captain Robert Singh (not uncoincidentally, another Indian) reflects on the religious zealots. "Now that he was forced to think about the previously unthinkable, it was not so astonishing after all. Almost every decade, right through human history, self-proclaimed prophets had predicted that the world would come to an end on some given date. What *was* astonishing—and made one despair for the sanity of the species—was that they usually collected thousands of adherents, who sold all their no-longer-needed possessions, and waited at some appointed place to be taken up to heaven. Though many of the 'Millennialists' had been imposters, most had sincerely believed their own predictions. And if they had possessed the power, could it be doubted that, if God had failed to cooperate, they would have arranged a self-fulfilling prophecy?"[9]

Clarke would have been aware of the fact that there was another book called *The Hammer of God* that, ironically, had been written by one of the religionists he scorned and made villainous in his own book. It was authored by a Swedish Lutheran bishop named Bo Giertz and was published in 1960 as a defense of the Gospel, which is to say unwavering orthodox Christian faith against the inroads of the dreaded liberals and their free-thinking, which he considered deplorable. He argued for the absolute power of God's word over spiritual deadness and rationalism, which he thought was a handicap because it is contrary to the sheer joy and salvation that come with faith. It is actually faith, not rationality, that is liberating. Giertz's *The Hammer of God* took its title from Jeremiah 23:29, "Is not my word like a fire? Says the Lord. And like a hammer that breaks the rock into pieces?"[10] Giertz would be all but deified by The Reborn.

"Sooner or later, it was bound to happen," Clarke wrote in the opening of *Rendezvous with Rama*, in a reference to the explosion over Tunguska (which either gave him the idea to write the book or reinforced it). It was another clever use of fact to fortify the fiction, which made the fiction more believable, as is the fictionalization of real competition between good and evil on the Western frontier (i.e., lawmen like Wyatt Earp and Bat Masterson against outlaws like Billy the Kid, the James brothers and the Daltons), wrestling alligators, charming snakes, landing at Normandy, trying to take Mount Suribachi from its hara-kiri-crazed defenders, and other dangerous and competitive endeavors. But the Hammer, of course, was the ultimate, consummate evil (and therefore, ironically, a force that could end evil as well as good).

"On June 30, 1908, Moscow escaped destruction by three hours and 4,000 kilometers—a margin invisibly small by the standards of the universe," Clarke wrote. "On February 12, 1947, another Russian city had a still narrower escape, when the second great meteorite of the twentieth century detonated less than four hundred kilometers from Vladivostok, with an explosion rivaling that of the newly invented uranium bomb." A generation that had seen photographs of the grotesque results of what manmade nuclear impactors did to Hiroshima and Nagasaki got the message. "In those days there was nothing that men could do to protect themselves against the last random shots in the cosmic bombardment that had once scarred the face of the Moon."[11]

Again, Clarke's genius lay in combining fact and fiction, as he did when he mentioned SETI (the Search for Extraterrestrial

Intelligence), and in basing his fiction on fact and clearly describing both, as he did in a chapter in *The Hammer of God* called "Excalibur," which accurately describes the nuclear-pumped x-ray laser that was designed for President Ronald W. Reagan's Strategic Defense Initiative ballistic-missile defense system, better known as Star Wars. (More about SDI in chapter 7.) Although he had a degree in mathematics and physics from King's College in London, which he received after he served as a radar specialist in the Royal Air Force in World War II, Clarke's extensive knowledge of the physical sciences, rocketry, and the whole realm of space was essentially self-taught. "My involvement with the subject of asteroid impacts is now beginning to resemble a DNA molecule: the strands of fact and fiction are becoming inextricably entwined," he explained with impressive candor.[12]

Since Clarke knew Luis Alvarez before the discovery at Chicxulub, followed that earth-shaking event closely, and kept up with the asteroid and comet situation, he would have known about the work in that area that was going on at universities and in the foundations and societies. He was also aware of the three-stage planetary-defense strategy against potentially hazardous space objects that NASA, the European Space Agency, Roskosmos, and just about everyone else believed and continues to believe will work. It is based on the B612 Foundation model, which Clarke used in *The Hammer of God*. The first stage is spotting Kali, taking its measurements, and determining that it is headed this way and is big enough to destroy a large part of the planet or perhaps all of it. Then *Goliath* is sent to nudge Kali off course and, if that does not work, to take it out, as the fastidious call it when obliterating something.

So Kali has to be sized up in *The Hammer of God*, as the real strategy requires.

The mass of Kali was known to within one percent, and the velocity it would have when meeting Earth was known to twelve decimal places. Any schoolboy could work out the resulting half em vee squared of energy—and convert it into megatons of explosive. The result was an unimaginable two million *million* tons—a figure that was still meaningless when expressed as a billion times the bomb that destroyed Hiroshima. And the great unknown in the equation, upon which millions of lives might depend, was the point of impact. The closer Kali approached, the smaller the margin of error, but until a few days before encounter, ground zero could not be pinned down to within better than a thousand kilometers—an estimate that many thought was worse than useless.[13]

Goliath's crew plants on Kali the most powerful bomb ever made. It doesn't explode as planned, but it does go off with enough force to cause the porous, moonlike rock to "fission like an amoeba." Both halves miss Earth.

And by way of showing that there is not just one death-dealing monster out there but many, Clarke introduced another one toward the end of the book.

> Kali 2 entered the atmosphere just before sunrise, a hundred kilo-meters above Hawaii. Instantly, the gigantic fireball brought a false dawn to the Pacific, awakening the wildlife on its myriad islands. But few humans; not many were asleep this night of nights, except those who had sought the oblivion of drugs.
>
> Over New Zealand the heat of the orbiting furnace ignited forests and melted the snow on mountaintops, triggering ava-lanches into the valleys beneath. By great good fortune, the main thermal impact was on the Antarctic—the one continent that could best absorb it. Even Kali could not strip away all the kilometers of polar ice, but the Great Thaw would change coastlines all around the world.
>
> No one who survived hearing it could ever describe the sound of Kali's passage. . . .[14]

And he used a suitably dramatic word for what Kali and other killer asteroids were capable of committing; a name that perfectly suited a world-ending cataclysmic act of violence: *terracide*.

Clarke could not resist beginning the novel by having some

fun with himself, some self-deprecation, which showed impressive self-assurance and respect:

> SPACEGUARD had been one of the last projects of the legendary NASA, back at the close of the Twentieth Century. Its initial objective had been modest enough: to make as complete a survey as possible of the asteroids and comets that crossed the orbit of Earth, and to determine if any were a potential threat. The project's name—taken from an obscure Twentieth Century science fiction novel—was somewhat misleading; critics were fond of pointing out that "Spacewatch" or "Spacewarn" would have been much more appropriate.[15]

The name Spaceguard, which would be used for the Spaceguard Survey project ordered by Congress five years after *The Hammer of God* was published, and which has now been adopted by the whole planetary-defense establishment (like the Spaceguard Foundation, for example) was first mentioned in *Rendezvous with Rama*, the "obscure" novel to which Clarke referred. It takes an artist with substantial self-confidence to refer to his own work as obscure.

Clarke was generally appreciated by the intellectual establishment not for being an outstanding writer but for very effectively combining fact and fiction in interesting, informative, and visionary works. "As much as any living writer, Clarke represents both the great imaginative strengths and the traditional literary limitations of classic science fiction," Gary K. Wolfe, a professor of English and humanities at Roosevelt University, wrote in the *New York Times*. "One does not read his work for depth of character or intricacy of plot, and his style is often pedestrian—though marked by clarity and capable of rising toward a resonant cosmic poetry when the subject is grand enough. But his field of vision is immense, filled with wonder and yet anchored by reason and meticulous detail. At its best, his work echoes the same fundamental questions that so transformed the hominids in *2001* [the space-odyssey film], and that strike to the sources of both myth and science."[16]

"As an asteroid named 'Kali' hurtles toward earth on a collision course that spells the end of life on the planet, a lone spaceship armed with a weapon to alter the asteroid's path attempts to carry out its perilous mission—unaware that others are simultaneously working for earth's destruction," *Library Journal* reported. "In the capable hands of science fiction veteran Clarke, a standard cosmic disaster plot becomes a lucid commentary on humanity's place in the cosmos. A good choice for science fiction collections."[17]

"Just to declare my interest up front, I'm a professional astronomer who observes comets and asteroids and has observed quite a few asteroids of the type that could impact the Earth," wrote one expert who found the book scientifically lacking. "I've read a number of books that use well-aimed comets and asteroids to bring universal doom—it's a subject which has been well-exploited in the last few years. Some books, like *Lucifer's Hammer* . . . are far superior in detail, although set in the present, rather than Clarke's far future. Compared to some of the books that I have read, the *Hammer of God* was disappointingly lightweight. What I will acknowledge is the future setting which Arthur C. Clarke invents and which is far more interesting and realistic in many senses than the Earth-impact part of the plot."[18]

Lay readers did not see it quite that way. "I loved this book," one of them wrote to Amazon. "Clarke takes a different approach in *Hammer of God*, switching from technological forecasting to sociological brainstorming, and hits one out of the ballpark. His predictions are hilarious. Christianity and Islam merge into a single religion [Chrislam]! On the surface his ideas seem absurd, but a quick glance at the front page of your daily newspaper suggests that Clarke's ideas may be closer to reality than one would like to think. The only thing about *Hammer of God* I didn't like was that it was too short! I know Earth will be saved, but Clarke creates such an interesting social panorama that I want to know more."[19]

Clarke expressed a more sanguine view of asteroids in *The Exploration of Space*, a short, nonfiction primer on space travel that was published in 1951. "Whether they will be of any interest to astronautics, only the future can tell. Although so many thousands of them exist, they cannot constitute a 'menace to navigation,' as has sometimes been suggested. The gulf between Mars and Jupiter is too enormous for a few thousand, or even a few million, asteroids to go very far towards filling it."[20] Not a menace to navigation, meaning space travel, perhaps, but surely a menace to Earth, which he certainly knew to be the case.

Those who want to be entertained by the specter of the end of life, at least on this planet, and learn about it have fostered a small but somewhat profitable genre: Doomsday (or "Doom$day") productions. They come to Meteor Crater in Arizona and are usually somewhat awed by what they see. They are actually looking at—witnessing—what a projectile, and not a very big one at that, can do to its target and, by implication, to the people on it.

Since he was (pardon) grounded in the physics, Neil deGrasse Tyson was awed by his first trip to the crater because it turned theory into reality. "The juxtaposition of appearance with accurate knowledge can be the most humbling force on the human soul. On first sight, the crater is simply an enormous hole in the ground—fourteen football fields across and deep enough to bury a sixty-story building," he recalled. "With the Grand Canyon a few hundred miles away, Arizona is no stranger to holes in the ground. But to carve the Grand Canyon, Earth required millions of years. To excavate Meteor Crater, the universe, using a sixty-thousand-ton asteroid traveling upward of twenty miles per second, required a fraction of a second. No offense to Grand Canyon lovers, but for my money, Meteor Crater is the most amazing natural landmark in the world. The polite (and scientifically accurate) word for asteroid impacts is 'accretion.' I happen to prefer 'species-killing, ecosystem-destroying event.'"[21]

Visitors soon learn that the site of the event is also called Barringer Crater. That name's origin goes back to 1902, when a successful Philadelphia mining engineer named Daniel M. Barringer heard about the crater, the meteor impact theory, and the chunks of iron that were said to be buried all around the site. He quickly became convinced that the crater and the area immediately surrounding it contained a mother lode of iron ore that was worth a fortune, so he formed the Standard Iron Company, bought the crater and the land around it, spent twenty-seven years looking for the metal at a cost of more than $800,000 ($10 million today) and ended up with neither iron nor money.[22]

Barringer's descendants were, and remain, imaginative individuals who figured out that mining iron wasn't the only way to make a living off of the crater. They claim that it is "the first proven, best preserved meteorite crater on earth," and sell tickets for admission to see it. Since many visitors want keepsakes to remind them of what could happen to them on the ultimate bad day, there are plenty of souvenirs. Authentic Meteor Crater dust can be taken home and displayed in the living room as a reminder of how fortunate its owners' ancestors were to have survived that impact and all the others. There are also Meteor Crater magic eggs, impact caps, special shot glasses and playing cards, mouse pads, a videotape about collisions with things that go bump in the night (and day) and that caused the crater, and genuine, freeze-dried Alien Ice Cream (a snack that gives new meaning to a taste that is out of this world). The ice-cream wrapper shows a smiling visitor from very far away enjoying a refreshing frozen treat after its long journey. It looks like Caspar the Unfriendly Ghost.

And there is a seventy-nine-page booklet, *The Meteor Crater Story*, which opens with a grabber:

Fifty thousand years ago, a giant invader from outer space hurtled through our Earth's atmosphere at incredible velocity and collided with northern Arizona's rocky high plateau. The meteorite's explosive impact destroyed all living things within a radius of several miles, created the chasm we call Meteor Crater, and strewed rock and meteorite fragments across a wide area.

Should an object of similar size and velocity strike New York City or any other densely populated area today, it could kill some ten million people.[23]

Barringer's role in the history of the crater is explained, including as the founder, as the president, and as a major financial contributor to the Meteoritical Society, a group of scientists that began in 1933 to study "astrophysical science." According to the booklet, Dr. F. C. Leonard delivered the opening address at the society's first meeting and predicted that meteoritics would one day enjoy "a definite and creditable standing among the sciences."[24]

"Could it happen again?" *The Meteor Crater Story* asks, fittingly, at the end. Sure. But, like most other respectable tracts on the subject, the booklet claims that it is a matter of degree. "True, it is not likely to happen. Neither is winning a multistate lottery. But someone wins a $50 to $100 million jackpot every few weeks."[25] An appendix lists 148 impact sites around the world, which is like ending with a large exclamation mark.

The entertainment world was also quick to realize that the proven potential of large rocks and comets to cause terrible damage is inherently dramatic and therefore financially profitable; that people will pay for the relief of seeing mass death and destruction from which they are mercifully spared as they paid to see monster movies like *King Kong*, *Frankenstein*, and *The Thing* and watch Bela Lugosi suck the blood out of other people and turn them into vampires as Count Dracula in that creepy haunted castle in Transylvania when he wasn't sleeping in his coffin. (Moral: always carry a crucifix in Transylvania.)

The film *When Worlds Collide* came out in 1951 and was based on a novel written by Philip Gordon Wylie and Edwin

Balmer that was published in 1933 (two years after Universal Pictures set Dracula loose). It is about a South African astronomer who discovers a star named Bellus that is on a collision course with Earth. He secretly sends copies of the radar imagery to a colleague in the United States—there would no doubt be a worldwide panic and anarchy if word got out that a civilization-ender was headed this way—who warns the United Nations delegates that their planet is only eight months away from a catastrophe. The astronomer pleads for the construction of a large spaceship that will take a lucky few to a planet called Zyra that is orbiting Bellus and that will therefore pass very close to Earth. Neither the world body nor the US government believes him, so he gets private funding from a wealthy, crippled, cantankerous industrialist and has the vessel built. Meanwhile, other scientists begin to agree that a cosmic collision is indeed imminent. Most of the spaceship's passengers are chosen by lottery, though a privileged few are selected by Dr. Cole Hendron, a distinguished and well-known scientist. (Sydney Boehm, who wrote the screenplay, had sense enough to avoid any reference to Noah.) As Doomsday approaches, the spaceship is stocked with food, medicine, animals, and as much of civilization's record as can be loaded on board. (It is not mentioned, but it is clear that the survival strategists do not want what happened to the Great Library at Alexandria to happen again.) The multitude of terrified people who lose the lottery riot and fight to get on board, but they are kept off by use of force, and the ship departs with its vitally important human specimens. Once in space, they see television pictures of the place they left being pulled into Bellus's relentless gravity, breaking apart, and finally exploding. The survivors, which formula required include two lovers, one a handsome, athletic male, and the other a beautiful female—for breeding purposes—land on Zyra in due course, find it habitable, and settle down on their new world.

Deep Impact, which was almost undoubtedly conceived by someone who had read Walter Alvarez's account of the Chicxulub expedition, was made by Paramount Pictures and DreamWorks and was released in 1998, the year of the NEO awakening and the congressional mandate to conduct the Spaceguard Survey. It grossed more than $349 million world-wide on an $80 million production budget, which put it in the financial winner's circle.

The plot has to do with a very large comet that is discovered by a teenage astronomer in an astronomy club in Richmond, Virginia. His teacher alerts an astronomer named Marcus Wolf who tries to get word out but is killed in an automobile accident before he can do it. A year passes. Then a television news reporter investigating the resignation of the secretary of the treasury finds a mysterious reference by him to "Ellie," who she thinks is his mistress. But a little more digging shows that Ellie stands for "E. L. E.," which, in turn, is an acronym for "Extinction-Level Event." Since the comet is eleven kilometers long, there will indeed be extinction if it reaches Earth. The story is aired.

Meanwhile, if politics does indeed make strange bedfellows, so does the distinct possibility that the world is about to end. The United States and Russia have therefore secretly built a spacecraft in orbit that is supposed to carry astronauts and cosmonauts to the approaching monster and blast it to smithereens. Capt. Spurgeon "Fish" Tanner, played by Robert Duvall, leads a team that lands on the fast-approaching comet and plants the standard weapon of choice, which is supposed to force the thing off course. But, confound it, the supernuke instead blows it into two huge pieces, both of which head straight for the Western Hemisphere.

While a last-ditch, unsuccessful effort is made to stop the two speeding impactors with nuclear missiles, the president declares martial law; two hundred thousand preselected scientists, teachers, engineers, artists, and soldiers are given special emergency protection; and a national lottery is conducted for ordinary folks to see who will be given shelter in limestone caves in Missouri and who will be reduced to their component parts or turned into steam. While individual sagas are playing out among the astronauts and their loved ones, one of the comet halves lands in the Atlantic Ocean off North Carolina, causing the mother of all tsunamis, which kills millions. But the other half of the comet is blown apart with the suddenly invaluable H-weapons by Fish Tanner and his crew after they decide to take on a suicide mission. Earth is spared yet again.

"Apparently there is no better aid to family therapy than a murderously large meteor hurtling toward Earth," *New York Times* film critic Janet Maslin wrote in her appraisal of the film.

> So the costly comet thriller *Deep Impact*, which is to summer movies what the first crocus is to springtime, explores the salutary effects of imminent doom. Lovers bond, family ties bind and old wounds heal as the planet prepares for its final hours, although the crisis proves not as dire as it could have been. We will survive to be hit by another comet picture (*Armageddon*) in July.
>
> *Deep Impact* will doubtless seem the more sensitive of the two, since it emphasizes feelings over firepower whenever possible. Mimi Leder, who directed *The Peacemaker* and gives greater gloss and personality to this film, directs with a distinct womanly touch. Within the end-of-the-world action genre, it's rare to find attention paid to rescuing art, antiques, elephants and flamingos.[26]

The villain in *Armageddon* is a "rogue" comet that, while passing through the asteroid belt, pushes an asteroid the size of Texas and other large rocks toward Earth. NASA becomes aware of the impending bombardment when a massive meteor shower clobbers the East Coast, including New York, and Finland and Shanghai. Of relatively little consequence, one of the comet pieces turns the orbiting shuttle *Atlantis* into a cloud of metal fragments. The astronomers yet again sound a timely warning by declaring that the Lone Star asteroid is due to hit Earth in eighteen days.

Given the damage that would be caused by a relatively puny kilometer-size asteroid, one the size of Texas would, at minimum, cause another extinction event like the one that started the long night and therefore did in *Tyrannosaurus rex*, its cold-blooded relatives, and the vegetation they lived on. At worst, it would disintegrate the planet, or dematerialize it, as NASA's scientists might put it. They hurriedly came up with a plan to blow the rock in two with a nuke so that both halves would separate and safely pass Earth. Then, however, a chunk of the comet that could be the size of Houston or Fort Worth (to push the metaphor to its limit) pulverizes Paris. Two shuttle orbiters, *Freedom* and *Independence*, are assigned to land on the next approaching asteroid, plant a remotely controlled nuclear bomb on it, and leave. But *Independence's* hull is punctured by one of the cometary fragments as it approaches the asteroid and it crashes, leaving *Freedom* to make a safe landing on it so the bomb can be planted.

The asteroid heats up as it gets closer to Earth, though, and that causes a rock storm that damages the bomb's trigger so that it cannot be set off remotely. Then, with Earth hanging in the proverbial balance, a command decision is made. One person is going to have to land on the asteroid and manually detonate the nuke, meaning that he will commit suicide to save the whole planet and all of humanity.

That man is Harry S. Stamper, the world's best deep-sea oil driller (and, thus, an explosives expert) played by Bruce Willis, who is left on the asteroid by *Freedom* and proceeds to do his immortal duty by manually setting off the buried "device" and martyring himself. *Armageddon* fared better at the box office than it did with reviewers, another indication that the public remains acutely interested in its fragile home's survival in a hostile and dangerous environment. It brought in $554,600,000 with a production cost of $140,000,000, making it the highest grossing film worldwide in 1998. And it was nominated for four Oscars, all of them for technical achievement such as best visual effects and best song.

It should be noted that nominations, as opposed to awards, do not count for much. Any work, however lacking in substance or creativity, can be nominated for an award, including Nobels, Pulitzers, and other prestigious prizes. It is winning that counts, and *Armageddon* had the ignominious distinction of winning a Golden Raspberry Award (also known as the "Razzies") for worst actor, Bruce Willis.

Maslin skewered it. Here she is, winding up for the pitch:

> Doom threatens. Again. This time it's a giant asteroid. ("It's the size of Texas, Mr. President"), and it's the Chrysler Building that becomes New York's most conspicuously flattened landmark (just as *Deep Impact* toppled the Statue of Liberty and *Godzilla* wrecked the Brooklyn Bridge.) That damage is done by a fake meteor shower during the first part of *Armageddon*. The sight, however apocalyptic, isn't as scary as the prospect of raising a generation of Americans on movies like this. Movie isn't actually the best word to describe *Armageddon*. More accurately it's a product, a feat of salesmanship, a sight worth noticing only because, like the asteroid on a collision course with planet Earth, its size and inevitability aren't easy to miss. But it should surprise no one to learn that the catchy title and prime opening date were more vital to the genesis of *Armageddon* than the burning need to tell one more derivative disaster story. . . . Though it means to be inspiring, it has quite the opposite effect. There's not a believable moment here (unless you count some boyish carousing in a strip club). The actors mark time, and the gung-ho heroics on display are embarrassingly hollow. . . . A real movie about courage in space is *Apollo 13*, in which fear and sacrifice have meaning. This jingoistic, overblown spectacle is about whistling in the dark.[27]

Michael O'Sullivan reviewed the film for the *Washington Post*:

Like a white-water ride on Class V rapid, *Armageddon* is a loud, long and bumpy experience. It might make you tense, it might make you nauseous, and its clangorous roar could well give you a migraine headache. Then again, when it's all over you might just want to throw up in a bucket, buy another ticket and get back in the boat for a second adrenaline-stoked slide down that swollen stream. Allow a day for recovery, however, because the nearly three-hour film is emotionally and physically exhausting. It's an intensely visceral pleasure, not unmixed with pain, like the multiple g-force acceleration experienced by an astronaut during lift-off. *Armageddon* peels your eyelids back and blows your eardrums out until rational analysis is moot. . . . But the special effects are stupendous and the suspense is palpable. By the film's end . . . you may resent the fact that every imaginable button of yours has been pushed raw, but you will be powerless to lift a finger to stop it.[28]

"Bruce Willis saves the world," Todd McCarthy opined in *Variety*, "but can't save *Armageddon*. The second and, mercifully, last of the season's nuke-the-asteroid-or-bust pre-millennium spectaculars is so effects-obsessed and dramatically be-numbed as to make *Deep Impact* look like a humanistic masterpiece. Despite its frequently incoherent staging and an editing style that amounts to a two and a half-hour sensory pummeling, $150 million sci-fi actioner nonetheless the Willis juice, Jerry Bruckheimer–Michael Bay bad-boy ingredients and Disney marketing muscle going for it to launch it into high commercial orbit."[29] (Bruckheimer, Hurd, and Bay were its producers.)

Jeanine Basinger, a film scholar, thought otherwise and called the film a "work of art by a cutting-edge artist who is a master of movement, light, color, and shape—and also of chaos, razzle-dazzle, and explosion. The film makes these ordinary men noble, lifting their efforts up into an epic event. If that isn't screenwriting, I don't know what is."[30] It was a complement not only for Bay but for herself, since she was his teacher at Wesleyan University.

Armageddon did not win plaudits for scientific accuracy, either. Bay admitted in an interview with *Entertainment Weekly* that the film's central premise—that the space agency could blow a large, Earth-approaching asteroid in half—was unrealistic. NASA, which was quietly trying to convince Congress and the Clinton administration that planetary defense should be one of its major programs, provided technical assistance to Disney Studios, but it was careful to explain that cooperation with *Armageddon*'s makers in no way indicated that it believed the movie's premise was scientifically plausible. Leaving nothing to chance, the space agency had a disclaimer inserted near the end of the credits stating, "The National Aeronautics and Space Administration's cooperation and assistance does not reflect an endorsement of the contents of the film or the treatment of the characters depicted therein."[31] And for good measure, it awarded the film a left-handed compliment by showing it as part of a management training program in which prospective managers were asked to find as many inaccuracies in the movie as they could. The total came to 168.[32] Bay reacted to the scorn by apologizing for making *Armageddon* in only sixteen weeks, which, he indicated, did not give it as much time as it deserved.

The international physics community apparently thought that the film was worth no time at all. The physicists said that production time was not its problem. An article called "Could Bruce Willis Save the World?" in the *Journal of Physics Special Topics*, written by four members of the Department of Physics and Astronomy at the University of Leicester, was among several that took strong exception to the concept of nuking the colossal attacker to save Earth. In order for the plan to work, the authors maintained, a hydrogen bomb that is a billion times more powerful than the Soviet Union's "Big Ivan," the most powerful bomb ever detonated, would be required. Like the rest of the science community and almost everyone else, they suggested changing its course long before impact time. And they could not resist pointing out that poor Stamper was caught between a rock and a . . . nuclear weapon.[33]

But the negative reviews notwithstanding, the public continued to find films about objects that are too near Earth thrilling. That subject was guaranteed to fill theaters and attract television viewers who were not interested in international relations, particularly since peace had broken out between their country and a Union of Soviet Socialist Republics that had imploded and morphed into just plain Russia, with the likes of Mikhail S. Gorbachev proclaiming the start of perestroika and glasnost, a restructuring of the nation and openness, respectively. He was awarded the 1990 Nobel Peace Prize for ending the Cold War. A new threat was therefore in order, and if the Russians and Chinese would not provide it, the universe would. It was only natural.

Meteor was released in 1979, during the Cold War, and was an early example of how the threat to everyone could force even enemies to cooperate. It starred Sean Connery, playing Dr. Paul Bradley (not 007, for a change), an American scientist who has invented a secret orbiting-missile platform called Hercules to use against asteroids. Unfortunately, it has been turned into an orbiting superweapon for use against the Russians by expedient bureaucrats who are forever focused on relatively petty East–West military competition. The Russians have their own missile-carrying satellite that is named Peter the Great. But then an asteroid called Orpheus collides with a comet. (Whoever named the asteroid had a wry sense of humor, since Orpheus was the god of music, poetry, and philosophy in Greek myth, and was therefore no Kali. It was like naming a serial killer Howdy Doody.) A five-mile chunk of Orpheus, along with smaller fragments, breaks off and heads for Earth. The fragments arrive here first and cause terrible devastation, including to New York, which is mostly destroyed. (It turns out that the subway is a good place to hide, though.)

Bradley and his Russian counterpart, Alexei Dubov, meet to come up with a way to stop Orpheus, but the Russian denies that his country has its own missile platform in space. That changes after the first fragments have slammed into Earth, and both nations decide to cooperate or risk a collective outcome that the Communists definitely do not want. They fire three salvos of missiles at Orpheus, which finally explodes. That is more than can be said of the film, which was rated a dud by most of its cast and the reviewers, though it did have a cult following and influenced *Deep Impact* and *Armageddon*, both of which did considerably better in ticket sales and reviews. It also had an afterlife of sorts as yet another TV miniseries.

Meteor Apocalypse, which was released in 2009, was made by a company called The Asylum—it means refuge as well as an institution for the mentally unbalanced—and was about the world's nuclear nations getting together to fire their missiles at a comet that is heading across Earth's path. They succeed in hitting the thing, but pieces of it reach Earth (again), not only contaminating groundwater and sickening millions of people but also obliterating Los Angeles. Destroying that city showed a clear lack of civic loyalty since it is The Asylum's asylum. Then again, it was LA's turn, New York having taken its hit in *Meteor*.

A Fire in the Sky, which came out in 1978, was one of the first comet films made for television, and the villain's target was a relatively modest one compared to much of what followed: Phoenix, Arizona. Astronomers warned that the comet was bearing down on the hapless city, but no one believed them. That may have been at least part of the reason why graduate students in the University of Arizona's respected Department of Astronomy called the film "The Comet That Ate Phoenix."

Meteorites!, which made it to television twenty years later— not uncoincidentally at the time *Armageddon* and *Deep Impact* were released—was about a salvo of meteorites heading for Earth, and especially for Roswell, New Mexico. But the local officials ignore a warning because they don't want anything to interfere with the annual UFO festival, which is celebrated to commemorate what is alleged to have been a landing there by space aliens in July 1947 that left the wreckage of their space-craft on a ranch. In reality, a US Air Force investigation found that the wreckage was from a high-altitude surveillance balloon in a top-secret project called Mogul and published that finding in *The Roswell Report: Case Closed*. Since the locals turned the crash site into a profitable tourist attraction, they were delighted that the "government" denied that it was really a UFO, since that smacked of a conspiracy—the old cover-up—which increased the place's value. There is now an International UFO Museum and Research Center, complete with a library and life-size replicas of the large-headed creepy creatures. (And yes, one of them is green.)

Doomsday Rock, another television film, came out in 1997 and concerned a noted astronomer who figured out that a comet was headed this way based on an ancient civilization's time line. True to form, he is ignored by most people, but a few are convinced and, with him, take over a missile silo and use its tracking equipment to find the thing. Security at the silo, which would have been set up to stop an attack by armed Russian agents, is not up to stopping a small group of intellectual rabble.

Asteroid, yet another television nail-biter that ran the same year, had Dallas as the target and its very upset residents reacting in ways that just about covered the whole emotional spectrum. The common denominator of all of the impact flicks was humans reacting to overwhelming danger. That theme goes back to the *Iliad* and the *Odyssey* (although Odysseus would never have so much as thought of abandoning Ithaca to the Cyclops Polyphemus).

Astronomers are the sentinels, the ones in the planetary watchtower, who see possible catastrophe coming and heroically alert the world in much of the literature, in motion pictures, and on television, since they are the ones with the telescopes.

And unlike biologists, who create strains of deadly viruses in laboratories or find ways to kill them, and unlike physicists, who concoct terrible weapons or send people around the Solar System, astronomers' work is passive. They spend their time peering at the universe through telescopes, which is about as exciting—to those who do not understand the excitement of discovering new worlds or increasing our knowledge of ones that have already been discovered—as crocheting. That makes them the ideal people to become instant heroes—humanity's saviors—by necessarily being the first to see the potential world-ender coming.

In *Moonfall*, for example, science fiction writer Jack McDevitt has an amateur discover a new comet that is one hundred times the size of other comets and is traveling at ten times their speed, not toward Earth, for a change, but toward the Moon. That would be bad enough under ordinary circumstances, but it is infinitely worse because, when the discovery is made, no less a dignitary than the vice president of the United States is on the lunar surface about to inaugurate a just-completed moon base. Having been alerted, scientists study the situation and conclude that, in less than five days, the comet will crash into the Moon, shattering it into large fragments that will rain down on Earth, causing killer storms and other calamities. Hence the book's title: *moonfall*, as in rainfall and snowfall. The comet is finally nudged off course with barely hours to spare by intrepid spacemen and spacewomen.

If there were an award for the most bizarre and improbable end-of-Earth story, a two-part television series called *Impact*, which aired in Canada and in the United States in 2009, would be top contender. During the worst meteor shower in ten thousand years, a "rogue" asteroid hidden in the meteors hits the Moon with such force that it and part of the Moon are shot to Earth, where they penetrate the atmosphere and impact. Damage is relatively slight, so there is relief, but that turns out to be unwarranted because there are subtle effects, such as cellphone disruptions, odd tidal waves, and static discharges. The usual suspects—the world's leading scientists—soon conclude that the Moon itself has been changed because the asteroid that hit it was actually a brown dwarf, which is a failed star. That is, it is an object that does not have enough hydrogen mass to be a star but still manages to be something tangible. The dwarf embeds itself in the Moon with such force that the Moon is pushed out of its orbit and heads straight toward . . . (begins with an *E* . . .). The defensive unit has just thirty-nine days to get the fast-approaching monster satellite to change direction, or there will be a big splat (as Dana Mackenzie put it in his book of that name, which is about the Moon being created when an object larger than Mars hit the home planet in its formative stage with such force that it knocked off a chunk that went into permanent orbit). After an attempt to destroy the Moon with nuclear weapons fails, two of the scientists, an astronaut, and the de rigueur cosmonaut land on it, plant an electromagnetic Moon-busting device, and set it to go off after they leave. Two of them make it back to Earth, the Moon is blown in half yet again—thus saving Earth—and the cursed brown dwarf flies into the Sun, which is exactly what it deserves. The scientific community (which was overdosed on close-call fluff by that point), for the most part, ignored *Impact*, which is to say it had precious little impact on the people who really known about such things.

Cosmic collisions can provide fun and entertainment for the whole family. "Simulate the damage caused by comet and asteroid collisions with Impact: Earth!" one game maker advertises on the Internet, mentioning Chelyabinsk. "Impact: Earth! is an interactive tool that lets anyone calculate the damage a comet or asteroid would cause if it happened to collide with our planet. You can customize the size and speed of the incoming object, and then find out if mankind survives. (Usually it does.)"[34] WHEW!

The situation is perhaps best explained on a "human" level by Maksim Y. Nikulin, who owns a circus in Moscow where children can have their pictures taken sitting next to a full-grown Siberian tigress named Chanel. He claimed that the sessions are safe. Besides, he said, the appearance of danger is interesting and integral to the circus arts. "People go to the circus for adrenaline. . . . If it appeared to be entirely safe it would not be interesting."[35]

5

THE OTHER
SALVATION ARMY

What happened over Chelyabinsk may have startled the uninformed, but it came as no surprise to astronomers, other scientists, the international space community, and just about everyone who had learned about what happened to the dinosaurs and who took that seriously. The possibility that the worst-case "event"—Doomsday—could happen because of an asteroid or a comet impact, or even that a region could be devastated, had long since gotten governments' attention and that of many ordinary people around the world who coalesced into groups dedicated to a new concept: planetary defense. Two generations that had thought a third world war, inevitably thermonuclear, would be the worst possible occurrence to be inflicted on humanity began to see and viscerally understand that nature could provide an even worse one, and not just to humanity, but to all of Earth and, therefore, to every living thing on it.

It is hard to overestimate how important the discovery at Chicxulub was. Everyone who had taken Astronomy 101 and learned about the potential world-ending violence in the universe, or who had visited a planetarium like the Hayden in New York and had gotten to see and touch a cold, hard, scarred meteor or a fragment of one, could sense what horrific destruction a large-enough impactor could cause. They could sense the carnage emotionally but could not imagine it clearly since it would be so massive that it was beyond clear conception. And catastrophic impacts, real and potential, were by no means only ancient history, relegated to *Tyrannosaurus rex*, *Stegosaurus*, *Allosaurus*, *Pterodactylus*, and the other creatures that disappeared when the light went out. The morning newspaper and the evening news carried stories of attacks on the planet, as they had for years. It was only a matter of noticing the evidence and establishing a pattern for the bombardment; of connecting the proverbial dots.

A "meteor shower" started over New York City on November 15, 1859, and lasted well into 1860 as fascinated New Yorkers watched the spectacular, panoramic sky show. On August 13, 1930, three meteors exploded over the upper Amazon with such force that hundreds of miles of jungle were set ablaze and continued to burn for months, covering the surviving trees in a vast area with a blanket of white ash and sending indigenous tribes fleeing in terror. On June 24, 1938, one estimated to have weighed 450 tons blew up roughly twelve miles above Chicora, Pennsylvania, and caused one apparent fatality: a cow. On August 3, 1963, a large bolide, which is loosely defined as a fireball, was detected 1,100 kilometers west-southwest of the Prince Edward Islands. On September 17, 1966, it was Lake Huron's turn when something exploded with a force equivalent to 0.6 kilotons of TNT over that body of water, which separates the state of Michigan and the province of Ontario, Canada. The Canadians got another one about five months later when a second meteor blew up thirteen kilometers above Alberta. The so-called Marshall Islands Fireball on February 1, 1994, was a double explosion that occurred with a force of eleven kilotons of TNT. That one was seen by an ambiguously named US military Defense Support Program 647 satellite, whose infrared eyes were designed to spot Soviet ballistic-missile tests and the hundreds of simultaneous hot spots that would have signaled an all-out missile attack and, therefore, the start of World War III.[1] There was an airburst caused by three breakups 150 kilometers south of Greenland on December 12, 1997.

Walt Whitman was so taken by comets and that long meteor shower over New York that he rhapsodized about them in a poem he called "Year of Meteors (1859–60)" that was in *Leaves of Grass*:

> Nor the comet that came unannounced out of the north,
>   flaring in heaven;
> Nor the strange huge meteor procession, dazzling and clear,
>   shooting over our heads,
> (a moment, a moment long, it sailed its balls of unearthly light
>   over our heads,
> Then departed, dropt in the night, and was gone;)[2]

Giotto's fresco *The Star of Bethlehem* shows a comet that some say is Halley's and that Giotto apparently interpreted as having spiritually religious significance. Van Gogh's *Starry Night* is a depiction of rather violent heavenly motion that is very likely cometary, and Charles Piazzi Smyth, who was an astronomer, portrayed the Great Comet of 1843 in a painting of that name that had a long, elegant tail that stretched diagonally across the entire canvas. There is also a dramatic engraving of the same comet passing over Paris. But it was interpreted, as the appearance of comets often are, as having a sinister side, as well; of heralding impending disaster. That notion was spread by William Miller, a Bible-thumping, former Long Island farmer in Hampton, New York, who was a captain in the army during the War of 1812 and who fell off the back of a wagon and landed on his head. Then he began having delusions that Earth would be destroyed by fire in 1843, when Christ would appear and raise the believers before the planet was purified by fire. Miller was convinced that when the comet appeared, it would signal the beginning of the transformation. Prophet Miller had some fifty thousand true believers.[3] Arthur C. Clarke may very well have had the Millerites (as they were called) in mind when he described The Reborn's trying to sabotage the mission to head off Kali in *The Hammer of God*.

There were many others in the late nineteenth and early twentieth centuries, however, who did not find asteroids and comets romantic or the heralds of paradise everlasting. Religious or secular, they looked at the record of destruction and decided to organize to learn as much as possible about the phenomenon and, in so doing, perhaps prevent more devastation, at least on a large scale. And the religious among them believed, and continue to believe, that the supreme being who created this world wants it and its inhabitants to survive and thrive, not demolished in a colossal explosion or disappear in an all-consuming fire. They decided that the only rational way to address the danger was to understand it, confront it, and create a protective system that included private organizations, governments, and, eventually, an international alliance to protect the planet.

What happened over Tunguska was well known within the world science community and so was what occurred over Brazil in 1930 and elsewhere, including Chicora, British Columbia in March 1965, Lake Huron that same year, and Chelyabinsk, which took its first hit in April 1944. Those "events" ratified the discovery at Chicxulub, and common knowledge about what happened to the dinosaurs did indeed make the situation seem like Russian roulette, with five chambers in the revolver being empty and the sixth containing a bullet that could end it all forever. The risk that near-Earth objects (NEOs) pose is usually perceived as a function of both the culture and the science of human society. "NEOs have been understood differently throughout history," Luis Fernández Carril, a Mexican climatologist, has written. Every time an NEO is seen, "a different risk was posed, and throughout time that risk perception has evolved. It is not just a matter of scientific knowledge." The perception of risk is therefore "a product of religious belief, philosophic principles, scientific understanding, technological capabilities, and even economical resourcefulness."[4]

People did indeed bring their own beliefs, principles, and

scientific understanding to the subject. But the common denomi-nator was the amalgamation of knowledge about what did in the dinosaurs, what happened at Tunguska, the discovery at Yucatan and of other craters all over the place, and the sight-ings of flaming objects all over the world. That caused deep concern among many organizations, including the national space programs, and among informed individuals, many of whom coalesced into organizations that were specifically dedi-cated to addressing the impact threat and all its ramifications. It galvanized thousands of individuals who, out of concern for the long-term safety of their planet, expanded their worldview beyond politics, economics, and traditional international rela-tions. Preventing the end of the world and, in turn, the end of human existence —no great-grandchildren; no descendants at all; oblivion—became the ultimate, most consequential cause. In Samuel Johnson's immortal words, "Depend upon it, sir, when a man knows he is going to be hanged in a fortnight, it concentrates his mind wonderfully."

It certainly concentrated the National Aeronautics and Space Administration's collective mind. The threat comes from space, and as the nation's space agency, NASA had to meet that threat by taking on planetary defense. The glory days of the Apollo program were two decades old, its imaginative Skylab program was a fading memory, and Solar System exploration was drying up in the wake of the sensational Pioneer and Voyager missions of the 1970s. Its participation in constructing the International Space Station (ISS) starting in 1998 was a shared, distant, and somewhat embarrassing goal. What haunted NASA (but was never talked about in public) was the fact that the ISS was built as a long-duration training vehicle so that astronauts, cos-monauts, and spacemen and spacewomen from other nations could form an international partnership and practice for the eventual trip to Mars, which is a year each way, and have physi-cians, psychologists, and others see how they reacted physically, mentally, and emotionally to so long a stint in space. The Mars mission was always elusively theoretical—a brew of fact and fiction that was equal parts Wernher von Braun and Robert A. Heinlein. It was part of a grand program called Constellation in which astronauts were to return to the Moon and others were to head for Mars, which a succession of robotic space-craft, the latest being Curiosity, have carefully scouted. But with pressing needs on Earth and a budget deficit of $1.29 tril-lion (the second highest ever), President Barak Obama cancelled Constellation—aborted it, as they say in spaceworld—in 2010. Given the nation's pressing needs, including paying off wars in Iraq and Afghanistan and maintaining the armed forces (while warily watching China expand its sphere of military influence), sending people back to the Moon and then to Mars was seen as being frivolous. (For their part, China's leaders once proclaimed that they intended to send taikonauts to the Moon in an evident effort to demonstrate that their country was a superpower that was up to that task, but then they quietly abandoned the plan, no doubt because they decided that the price of the ticket was not worth the trip.)

There was nothing frivolous about the asteroid and comet threat, though. So that pivotal year—1998—NASA established the Near-Earth Object Program Office at the Jet Propulsion Laboratory (JPL) to coordinate the detection, tracking, and characterization (that is, the size, shape, weight, and compo-sition) of the potentially hazardous asteroids and comets that prowled the neighborhood. Behemoths that can cause extinc-tions are rare and tend to turn up roughly once a century, whereas most NEOs are relatively small. But so is a bomb that can demolish a house.

"These are objects that are difficult to detect because of their relatively small size but are large enough to cause global effects if one hit the Earth," explained Don Yeomans, the JPL astrophysicist who became the Near-Earth Object Program

Office's first director and who would write an informative and digestible primer on the situation, *Near-Earth Objects: Finding Them before They Find Us*, that ought to be required reading at every university on the planet. The book's importance extends beyond explaining the potential danger to lay people because, in addition to its technical responsibilities, the NEO Program Office is supposed to facilitate communication between the astronomers and other scientists who would have to deal with a potentially calamitous approaching asteroid or comet and a public that would range from being utterly ignorant and oblivious to being hysterical. Yeomans bridged that chasm with *Near-Earth Objects*, which amounts to a briefing manual that is very important for making clear what a collision or another very near miss would mean to Earth. The other mission NASA was ordered to take on in that fateful year of 1998, of course, was compiling the Spaceguard Survey—the catalog—within a decade, which it did.

A planetary "hanging" (to use that metaphor) concentrated Maj. Lindley N. Johnson's mind, too. Johnson was a US Air Force officer who invented the term *planetary defense* in a white paper he wrote in 1993 for the Air University SpaceCast 2020 study to determine what capabilities the air force should have in the post–Cold War world. The paper, *Preparing for Planetary Defense: Detection and Interception of Asteroids on Collision Course with Earth*, began with the assumption that a defensive system was definitely required and then made the point that the necessary technologies, including nuclear explosives, missile propulsion systems, guidance, and targeting, were effectively Cold War surplus and were therefore available. And since the threat is global, he added, the expense of operating the system should be shared by many nations. "The cost for such a system, which might be analogous to buying life insurance, also rightly belongs in the international arena," he wrote. The clear implication was that, given the consequences of a lack of defense against impactors, the expense would be inconsequential.[5] "What's your life worth?" was the unstated but clearly implied question.

Preparing for Planetary Defense described the problem at length, providing necessary background details on the objects, their destructive potential, and mentioning the usual suspects: the Tunguska and Chicxulub impactors and Shoemaker-Levy 9's attack on Jupiter. "Now that it is recognized that collisions with objects larger than a few hundred meters not only can threaten humanity on a global scale but have a finite probability of occurring, means for mitigating them seem clearly worth investigation," Johnson stated emphatically up front. "It should also be recognized that the technology required for a system to mitigate the most likely of impact scenarios is, with a little concerted effort, within humanity's grasp."[6] The technology consists of rockets carrying robots for distant intercepts that would deflect the approaching rock or comet—the old nudge-it-off-course-very-far-from-Earth technique—and a second requiring a forceful deflection with the use of "high energy options" that included nuclear and kinetic weapons that were then available and, within two decades, lasers and ultrahigh kinetic energy systems. And the paper looked to the distant future and mentioned antimatter weapons, mass drivers, solar sails that would use the Sun's radiant energy to gently push the rock off course, and, in a truly impressive leap of imagination, "asteroid eaters" that would infest it with devices that would replicate themselves by feeding on the asteroid itself. They would, in other words, cannibalize it. "Over the period of several months or a few years, these devices, recreating themselves into an army of thousands, could completely mine the asteroid away, or at least reduce it to a size that is no longer a threat or is more easily maneuvered by the propulsion technology." Pushing the asteroid off course, gently or with force, seems more practical and effective than staging a long-duration banquet. And Johnson makes no mention of excrement and what, if anything, to do with it. In any case, he did an enormous amount of homework, part of which included acquainting himself with *The Hammer of God*, which he was sure to mention.[7]

Two other air force officers, Lt. Col. Rosario Nici and 1st Lt. Douglas Kaupa, wrote "Planetary Defense: Department of Defense Cost for the Detection, Exploration, and Rendezvous Mission of Near-Earth Objects," which appeared in *Airpower Journal* in 1997. They did their homework, too. After describing the NEO situation and the requisite impact sites, including Meteor Crater and Jupiter's perforated atmosphere—but also Manicouagan crater in Quebec and Wolfe Creek Crater in Australia—Nici and Kaupa maintained that the US government, through the Department of Defense, "is obligated to protect the lives and safety of its citizens," and that the nation "may use its armed forces, under the hierarchy of interests, for cases of strict humanitarian concern. Thus, responding to the NEO threat could be seen to fall under this policy."[8] They noted that the chief of staff of the US Air Force had tasked Air Force Space Command to determine what would be needed for the defense of the planet by fiscal year 1997. Furthermore, an effective system had to include coordinated worldwide coverage of the sky, a determination of what sizes constitute progressively serious dangers, knowledge of the objects' composition to help mitigation strategy, rendezvousing with an NEO to study it up close, practicing attacking and destroying one, "capturing" and mining an asteroid, and setting up a system that would warn about "small" ones that could save lives and prevent tsunamis, earthquakes, and forest fires.

Nici and Kaupa also provided some cost guesstimates, which varied depending on the scope of the program, with $112 million a year for detection, exploration, and rendezvous over the course of twenty years being the minimum. "Assessing the NEO threat would be a small cost for insurance, whereas an impact would cost billions of lives and trillions of dollars," they concluded, yielding to the temptation to squander the obvious. "While there is no reason to fear NEOs daily, there is a finite probability another NEO will collide with Earth." It was the old "not if but when" refrain yet again.[9]

Given that the asteroid and comet threat is obviously international, the United Nations also got involved. In April 1995, the United Nations International Conference on Near-Earth Objects held a three-day meeting in New York that was organized by the United Nations Office for Outer Space Affairs for two immediate purposes: (1) to make member states aware of the problem—to sensitize them, as its organizers put it—and (2) to expand the international detection and tracking system. Scientists from around the world met to discuss the NEO situation and, specifically, to collect and interpret scientific data that would shed light on impact history, evaluate then-current observations, and outline exploration missions that it believed would have to be undertaken (that is, go out and look at the things up close).

A total of forty-six papers were presented in five general areas: astronomy, Earth and planetary sciences, astronautics, detection and mitigation, and UN-related issues. Astronomy, as was to be expected, described the situation, including what is out there. "Meteoroid Orbits: Implications for Near-Earth Object Search Programs," was one paper, for example, and "Long-Period Comets and the Oort Cloud" and "Comet Shoemaker-Levy 9 Fragment Size Estimates: How Big Was the Parent Body?" were two others. Earth and planetary sciences included papers on "Target Earth: Evidence for Large-Scale Impact Events" and "A Unified Theory of Impact Crises and Mass Extinctions: Quantitative Tests." Gerta Keller, the Princeton geoscientist who insisted that the impact at Chicxulub occurred some three hundred thousand years before the K-T boundary, made that point again in a paper that was wryly titled "Asteroid Impacts and Mass Extinctions—No Cause for Concern." Astronautics had but four papers, including "High-Performance Ultra-Light Nuclear Rockets for Near-Earth Objects Interaction Missions," while the detection-and-mitigation segment had eight papers, "Technology for the Detection of Near-Earth Objects" being a representative contribution. Four papers were delivered at the final meeting, which was about the United Nations' role, including "International Efforts toward the Spaceguard System" and "A Proposal to the United Nations Regarding the International Discovery Program of Near-Earth Asteroids." [10]

Six years later, with all of that information in hand, the UN Committee on the Peaceful Uses of Outer Space, or COPUOS, established the Action Team on Near-Earth Objects, or Action Team 14, as it was called. The "team" was mandated to review the content, structure, and organization of programs dedicated to planetary defense; identify gaps in the work that was going on, specify where additional international coordination was needed; note where other countries and organizations could make contributions; and propose steps for improving international coordination and collaboration with groups that had been formed specifically because of the NEO threat.

Action Team 14 issued its report on February 18, 2013—three days after the explosion over Chelyabinsk, which was an irony that Shakespeare would have savored. It recommended the formation of an international asteroid warning network, an impact disaster planning advisory group, and a space mission planning advisory group. The IAWN, as the first was called, would be a network of experts that would focus on the discovery, tracking, and observation of potentially dangerous NEOs. Once spotted, the objects' orbits would be predicted and any potential impact warnings would be sounded. The warning network would also prepare to go public with what it discovers while searching for more information and keeping COPUOS briefed. The IAWN was also supposed to report to the committee every year on the overall NEO situation. The Impact Disaster Planning Advisory Group, or IDPAG, would review what was learned from other major disasters and prepare coordinated response plans and exercises, or drills, to address predicted impacts and those that come as a surprise. And the Space Mission Planning Advisory Group, SMPAG (or "Same Page," as it was called), would combine the voluntary expertise of spacefaring nations' specialists and recommend and promote research on missions that would mitigate attacks on an international, cooperative basis and develop technical concepts and propose operational programs. That could be accomplished only if everyone was, indeed, on the same page.[11]

By that time, science publications were paying attention to developments in the NEO realm, and so were the news media. The vast majority of news organizations reported developments, including the occasional "event," like that explosion over the Eastern Mediterranean, accurately and unemotionally. The heyday of fireball sightings in New York was in the nineteenth century, and the *New York Times*, which was founded in 1851, covered many of them with a thoroughness and lack of sensationalism that became the self-styled newspaper of record's hallmark.

"This morning at 1:40 the most beautiful meteor seen in this vicinity for years flashed across the northern sky nearly from horizon to horizon," a stringer in Utica wrote in 1875, using an uncharacteristic superlative. "Lake-side cottage in this pleasant summer resort had a narrow escape from destruction by a meteor last night," read a dispatch from Schroon Lake, New York, in 1880.[12] Recognizing the inherent danger of large NEOs, the *Times*; other mainstream newspapers; and weekly magazines such as *Time*, *Newsweek*, and the *Economist* reported the explosion over Chelyabinsk extensively, as accurately as possible, and with no blatant hype (as it is called in the newsroom).

It is appropriate to quote the *Times* again. "Gym class came to a halt inside the Chelyabinsk Railway Institute, the students gathered around the window, gazing at the fat white contrail that arced its way across the morning sky. A missile? A comet? A few quiet moments passed. And then, with incredible force, the windows blew in," the Moscow bureau reported in a story that led the paper, meaning it "started" in column six, on the extreme right side of page one, then "jumped" to A8, where there was a picture of the meteor streaking across the sky and two "sidebars" that provided additional, supporting information. One of them, "A Flash in Russian Skies, as Inspiration for Fantasy," quoted Stephen Baxter, the president of the British Science Fiction Association, as saying, "I think we got overconfident in the 1990s" with movies like *Armageddon* and *Deep Impact*, "when we thought we could fend off any threat," he said. "H. G. Wells knew we couldn't."[13]

"The scenes from Chelyabinsk," the story continued, "rocked by an intense shock wave when a meteor hit the Earth's atmosphere Friday morning, offer a glimpse of an apocalyptic scenario that many have walked through mentally, and Hollywood has popularized, but scientists say has never before injured so many people."[14]

A. C. Charania and Agnieszka Lukaszczyk may have read the *Times*'s account of the mishap at Chelyabinsk and other accurate and evenhanded accounts of what happened, but they still expressed concern that the news media, which were commonly scorned for allegedly sensationalizing developments "to sell newspapers," would overplay such stories. Charania worked for SpaceWorks Engineering, and Lukaszczyk was a space-policy consultant and a member of the Secure World Foundation. In a paper titled *Assessment of Recent NEO Response Strategies for the United Nations*, which was a thoughtful plan for how the United Nations should handle the NEO situation, they succumbed to warning about the hysteria-prone news media.

"There is concern that there will be many warnings with an associated over-reaction by the media and subsequently the public at large (i.e., multiple Apophis scenarios). Some people have speculated that this may not be a desirable situation and that we may require a coordinated 'clearing house' to prevent such media excitement." They used the Large Hadron Collider, the world's largest and highest energy particle accelerator that is underground on the Swiss–French border near Geneva, to make their point, mentioning some early news accounts that it would create a mini black hole that would devour the planet. That was accurate as far as it went, but they neglected to distinguish between the responsible print media and the rags, with their respective television equivalents. And a clearing house would, to some extent, centralize information at one source, which would amount to a degree of control that news media in free societies tend to find dangerous.[15]

By then, the B612 Foundation, which came into being in October 2002, was playing a leading role in planetary defense. Mentioned earlier, in chapter 2, its name refers to the asteroid that Antoine de Saint-Exupéry's little prince called home, but like the asteroid named Orpheus, the organization's name has reverse meaning. B612 definitely does not want to turn an asteroid into home, sweet home. The foundation grew out of a one-day workshop on asteroid deflection held at the Johnson Space Center on October 20, 2001, and was (and remains) a private foundation dedicated to finding potential impactors so far ahead of collision that they can be nudged off course by so-called space tugs the way ocean liners and other large ships are nudged by the nautical variety. It was invented by JPL's Clark Chapman and others, most notably Russell L. "Rusty" Schweickart and Ed Lu, by then veteran astronauts of the Apollo 9 lunar mission and the STS-84 and STS-106 shuttle missions, respectively.

The foundation's core project is a solar orbiting infrared

telescope that is appropriately named Sentinel and is designed to locate and catalog 90 percent of asteroids that are 140 meters or larger in diameter so that any headed this way can be pushed in another direction long before a possible collision.[16] (The mission is described in detail in chapter 7: "The Ultimate Strategic Defense Initiative.")

The B612 Foundation is dedicated to "opening up the frontier of space exploration and protecting humanity from asteroid impacts," Lu explained in March 2013 at a Senate hearing that was called so the impact risk and preventive measures could be assessed. "On June 28, 2012, the Foundation announced its plans to carry out the first privately funded, launched, and operated interplanetary mission—an infrared space telescope to be placed in orbit around the Sun to discover, map, and track threatening asteroids whose orbits approach Earth."[17] That is Sentinel, a 7.6-meter infrared telescope that is being built by Ball Aerospace and Technologies Corp. in Boulder, Colorado. It is designed to orbit the Sun at the same distance as Venus and to look in this direction with the Sun effectively at its back all the time so its vision will be continuous and it won't need the astronomical equivalent of sunglasses. It is being designed to locate 90 percent of near-Earth asteroids larger than 140 meters and half of those larger than 50 meters. The total cost of building, launching, and operating Sentinel is expected to be about $400 million, which, given its mission, is absolutely inconsequential.[18]

Clark R. Chapman, a recognized expert on NEOs and coauthor with David Morrison of *Cosmic Catastrophes* (a readable rundown of all sorts of space-related life-threatening situations, including asteroid impacts), testified on the NEO impact threat before Congress in May 1998, when the danger was coming into sharp focus. He led off by dismissing impacts like the one that occurred in *Deep Impact* and that would create a new Dark Age as happening only once every one hundred million years and that the chance of it occurring in the twenty-first century are one in a million.

"A more serious problem, and one that we *can* do something about, is the chance that a smaller asteroid or comet, about a mile wide, might hit," Chapman said. "The best calculations are that such an impact could threaten the future of modern civilization. It could literally kill billions and send us back into the Dark Ages. Such an impact would make a crater twenty times the size of Meteor Crater in Arizona. The gaping hole in the ground would be bigger than all of Washington, D.C., and deeper than 20 Washington Monuments stacked on top of each other," he continued, no doubt getting the rapt attention of the members of Congress by raising the specter of where they were disintegrating in an explosion that would make terrorist attacks look like fireworks displays. The answer, Chapman said, echoing the revealed wisdom, is to first find out whether a mile-wide asteroid is headed this way. "We simply haven't been looking hard enough," he continued. "As the rates of discovery, of objects both large and small, goes up and the public becomes more aware of the danger from the skies, it will be essential that planetary protection be elevated from a sideline activity of a few astronomers, and some passionate amateurs, and be put on a sound, appropriately funded footing. The cost is not large. I believe that *Deep Impact* has already taken in more money at the box office than the cost of the entire Spaceguard Survey, from beginning to end. Astronomical programs are comparatively cheap. The really large expenses involve implementing mitigation hardware—rockets and bombs. Fortunately that won't be necessary until a threatening, mile-wide object is found to be headed toward Earth . . . and then, surely there will be no debate about using nuclear weapons in space—just once—to save civilization from catastrophe." And Chapman, too, saw fit to mention "the visionary science fiction writer Arthur C. Clarke."[19]

The Spaceguard Foundation, which takes its name from the visionary's planetary-defense system, is also dedicated to analyzing the danger and doing something about it. It was started in Rome in 1996 by Gene Shoemaker; Duncan Steel, an Australian astronomer and the author of *Rogue Asteroids and Doomsday Comets*; Andrea Carusi, an Italian astronomer; and the International Astronomical Union's Working Group on Near-Earth Objects. The foundation's stated goals are to promote "the protection of the Earth environment against the bombardment of objects of the Solar System (comets and asteroids)." Specifically, it is to coordinate the discovery and sizing up of NEOs internationally, study the mineralogical characteristics of minor bodies in the Solar System (particularly NEOs), and promote and coordinate a ground network (and possibly a satellite network) of radar and other sensor installations, which it called the Spaceguard System. It is based in Frascati, Italy, and is also private and nonprofit. There are other Spaceguard foundations or associations in Croatia, Germany, Japan, and the United Kingdom.

Spaceguard UK is also known as the International Spaceguard Information Centre, but that does not mean it does public relations in the sense that many other "information" outlets do it. The center is a working observatory near Knighton, in Wales. It became operational in 2001 after being founded by Sir Patrick Alfred Caldwell-Moore, a very accomplished, quintessentially eccentric, English character—a veritable institution, as his countrymen correctly put it. In fact, he began wearing a monocle at age sixteen, lied about his age to get into the Royal Air Force, developed an interest in astronomy when he was six and became a well-known amateur astronomer, was the author of more than seventy books on that subject, was the moderator of the BBC's long-running *The Sky at Night* series, became the president of the British Astronomical Association, was a self-taught glockenspiel and piano player, was a prodigious fiction writer, and was an implacable and outspoken opponent of fox hunting (which he called a "blood sport" where animals are killed for fun). Spaceguard UK is the hub of the Comet and Asteroid Information Network (CAIN), which coordinates information on NEOs with other organizations around the world, certainly including NASA and the European Space Agency (ESA).

ESA does not like to be thought of as tilting at windmills, but it nevertheless named its major project, two spacecraft on an asteroid collision mission, Don Quijote. The first spacecraft, Sancho, was supposed to orbit an asteroid for several months and study it. Then a second spacecraft, Hidalgo, was supposed to crash into it, whereupon Sancho was to look over the asteroid to see how the impact changed its shape, internal structure, orbit, and rotation. ESA had two possible target asteroids in its sights: one named Amor 2003 SM84, and the other 99942 Apophis (which is named after that evil ancient Egyptian serpent mentioned in an earlier chapter).

The Catalina Sky Survey (CSS) knows where Apophis and a lot of other asteroids are. It was created because of that 1998 congressional order to NASA to identify 90 percent of Near-Earth Asteroids a kilometer or larger (later reduced to 140 meters) within ten years. The survey got its name because it is in the Catalina Mountains north of Tucson, Arizona, which puts it within commuting distance of the University of Arizona's world-class Department of Astronomy. The inbreeding has produced impressive results. Using telescopes on two mountains in the area and a third in Australia for "down-under" coverage, the CSS has turned up more than 2,500 NEOs, including potentially dangerous asteroids and some comets. The Lincoln Laboratory at MIT runs the Lincoln Near-Earth Asteroid Research (LINEAR) program, which was started in 1998 and is funded by NASA and the US Air Force. It uses two electro-optical deep-space surveillance telescopes at the White Sands Missile Range in Socorro, New Mexico, where the Pentagon tested ballistic and other missiles during the Cold War. It had discovered 2,423 NEOs and 279 comets by September 2011.[20] Pan-STARRS, the Panoramic-Survey Telescope and Rapid Response System, operates on a mountain on Maui, Hawaii, is run by the University of Hawaii, and is largely funded by the air force. All of the above (and some others) report what they find to the Minor Planet Center—technically the Smithsonian Astrophysical Observatory—in Cambridge, Massachusetts, which is affiliated with the Harvard College Observatory and operates under the auspices of the International Astronomical Union.

The Planetary Society was founded in 1980 by Carl Sagan; Bruce Murray, a leading Caltech/JPL planetary scientist; Louis Friedman, another Caltech/JPL space explorer; and Bill Nye, who began his career as a mechanical engineer at Boeing and then found his calling and blossomed as an effervescent science educator, actor, writer, comedian, and talk-show host, and as "Bill Nye the Science Guy," the star of a Disney/PBS children's science show. The society encourages public membership and was created to encourage humankind's presence in space, Solar System exploration, the search for extraterrestrial life, and the study of NEOs. It supports NEO research by awarding annual Shoemaker grants that, in 2013, to take one example, came to $34,307 split among five winners. "As the Chelyabinsk impact demonstrated, asteroid impacts happen; they are dangerous, destructive, with no regards for human life," Nye said as the award recipients were announced at a planetary-defense conference in Flagstaff, Arizona. "Tonight we honor citizen scientists, amateur and professional astronomers who make tens to hundreds of thousands of follow-up NEO observations each year, and their work is the key to determining NEO orbits and protecting life on Earth."[21]

Like just about everyone else in the NEO club, the members of the Planetary Society have called for finding out what, exactly, the threat is, and they have suggested gaining access to US military surveillance satellite data—that's the black (for

secret) program run by the National Reconnaissance Office—
for small-scale impacts in the atmosphere. It also provides a
probability table on its website that was compiled by Clark R.
Chapman and David Morrison that shows the chances of an
individual dying from selected causes in the United States. A
motor-vehicle accident tops the list with 1 in 100, followed by
murder (1 in 300), fire (1 in 800), firearms accident (1 in 2,500),
passenger-aircraft accident (1 in 20,000), flood (1 in 30,000),
and then asteroid/comet impact (1 in 40,000).

The society parts company with most other NEO groups
by claiming that the comets, the asteroids, and their derivatives
are not (pardon) an unmitigated menace. It in effect says that
all the emphasis on asteroids being a menace ignores their positive aspects.

> Finally, it is vital to evaluate whether Near-Earth Objects really are
> our foes or our friends. Over the next three centuries, there is a 1 in
> 30 chance that a Tunguska-like impact will result in some human
> casualties and a 1 in 3,000 chance for a larger, global-scale impact.
> A Spaceguard survey, however, is certain to find in near-Earth orbits
> several thousand non-threatening objects that are more accessible
> than the Moon in terms of rocket propulsion. Over the next three
> centuries (and hopefully sooner), these objects can provide interme-
> diate mission destinations as we prepare for long-duration human
> flights to Mars. As we begin to utilize space, the metals and vola-
> tiles (chiefly water) we find in these objects may become vital space
> resources. Thus, in taking a long view of only a few centuries, it
> is most likely that we will know the Near-Earth Objects as our
> friends. The lesson for us now is to keep in mind that all friends
> need respect.[22]

The Planetary Defense Foundation is an international network of amateur and professional astronomers that calls itself
a company. Its purpose is to understand and protect Earth by
analyzing data based on the discovery of asteroids and comets
in the Solar System and to share the discovery of new NEOs
through a grid network that uses electronic imagery.

The Secure World Foundation was created to accomplish what its name says, and it therefore definitely does not see NEOs as beneficial. It is a rarity in the planetary-defense network: a family operation. It was started in 2002 by Marcel and Cynda Collins Arsenault, two exceedingly well-off Coloradans (he owned more than $200 million in commercial real estate alone in 2013) with a serious penchant for philanthropy who became obsessed with the goal of promoting a secure, sustainable, and peaceful environment in space for the stability of Earth. With the Cold War over, the Arsenaults decided that the time was right for an international effort to finally use all of space for the benefit of humanity; that it is an environment with an infinite capacity to help this civilization and the planet as a whole. He became the founder and president of the Arsenault Family Foundation and One Earth Future, and she, with forty years' experience in nonprofit work, including work in prisons, mental health, disability rights, Girl Scouts, 4-H, and environmental issues, became the foundation's chairman of the board.

The foundation's stated mission is to work with governments, industry, international organizations, and civil society to develop and promote ideas and actions for international collaboration that achieve the secure, sustainable, and peaceful uses of outer space. "As a global commons over which no country has sovereignty, outer space presents a particular challenge to the international community," they said in articulating the organization's challenges.

> The foundation holds the core belief that without international cooperation focused on creating appropriate institutional and legal mechanisms to govern behavior in outer space the world could suffer the well-known 'tragedy of the commons.' This refers to a dilemma in which multiple actors, working independently, and rationally consulting their own self-interest, ultimately deplete a shared limited resource even when it is clear that it is not in anyone's long-term interest for that to happen. Articulating measures to prevent the loss of use of outer space is one of the primary motivations

for forming Secure World Foundation. Cooperative and collaborative solutions for space sustainability and usability also provide increased interdependence and interconnectedness on Earth, which increases the world's security.[23]

And security means, among other things, defense against NEOs in what the foundation, echoing Shoemaker, called "a shooting gallery." The Arsenaults made it clear on their website that they understood the danger and the defensive requirements, but that did not interest them so much as "appropriate governance" and "facilitation and information sharing." The first had to do with how the world could organize to meet the challenge of mitigating the effects of an incoming impactor. "Planetary defense poses significant policy and legal challenges which echo some of the same problems found in other areas of the outer space realm. These include space situational awareness, data sharing, collective security, and shared decision making." The second, "facilitation and information sharing," had to do with being a source of information for individual nations, the international community as a whole, and the national space agencies to warn about the common danger. It also had to do with the promotion of dialogue and cooperation among all of them. "The Foundation has a strong interest in contributing to the important task of creating an internationally agreed upon plan and guidelines for responding to a NEO threat. Hence, the Foundation has partnered with the Association of Space Explorers and other organizations to assist the Committee on the Peaceful Uses of Outer Space to develop an appropriate international agreement for responding to the NEO threat."[24]

In common with the rest of the planetary-defense community, the Secure World Foundation was and remains emphatic about the danger being global and that it therefore requires a unified international response.

Kirill Benediktov heartily agreed. He was a bestselling Russian author, historian, and policy analyst who made a pre-

The meteor that exploded over Chelyabinsk, Russia, as seen in this rare photograph, injured more than 1,400 people and did extensive structural damage.

A view of the meteor before it exploded over Chelyabinsk. The long smoke trail led some in the area to believe it was an errant ballistic missile. (Photo courtesy of Konstantin Kudinov.)

After a long search, a chunk of the Chelyabinsk meteor was found and pulled out of Lake Chebarkul for detailed study. (Photo used by permission of Alexander Firsov for AP Photo.)

The nearly mile-wide Meteor Crater in Arizona, accessed by the long road in the foreground, attracts many tourists who are fascinated by Earth's being the target of a potential Doomsday impactor. As late as 1945, the US Geological Survey did not believe that the crater was made by an impactor because there are no fragments. (They had disintegrated.) Some thought that it is an extinct volcano. (Photo courtesy of Meteor Crater, Northern Arizona, USA.)

A realistic view of the Sentinel infrared telescope's position from which it is supposed to watch for threatening near-Earth objects. The telescope is in its orbit at the far left; the Sun is at the center, followed by Venus and, to the far right, Earth. (Image courtesy of Ball Aerospace and Technologies Corp.)

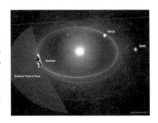

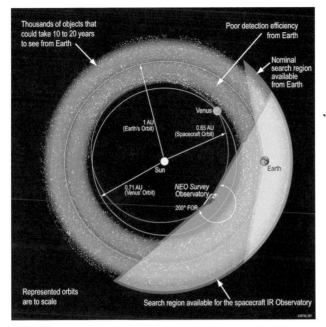

An overview diagram of the Sentinel infrared telescope's position in a Venus-like orbit from which it is supposed to watch the region around Earth for potentially dangerous asteroids and comets. (Image courtesy of Ball Aerospace and Technologies Corp.)

In a space first, the Japanese spacecraft Hayabusa (peregrine falcon) landed on this asteroid, 25143 Itokawa, in 2005, collected tiny samples, and returned them to Earth. (Photo © Japan Aerospace Exploration Agency [JAXA].)

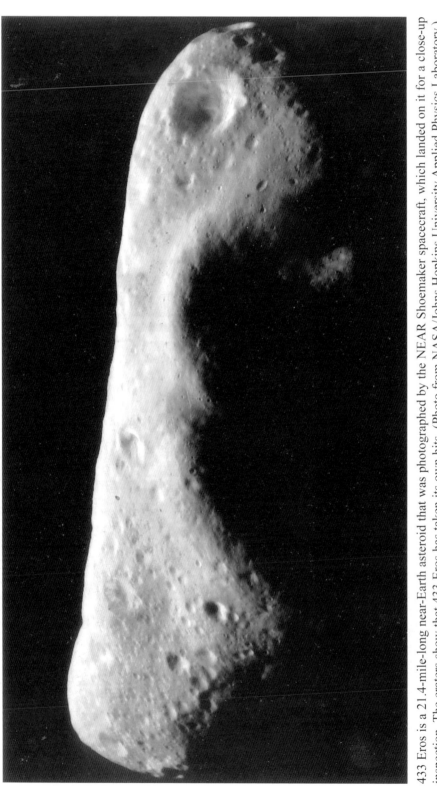

433 Eros is a 21.4-mile-long near-Earth asteroid that was photographed by the NEAR Shoemaker spacecraft, which landed on it for a close-up inspection. The craters show that 433 Eros has taken its own hits. (Photo from NASA/Johns Hopkins University Applied Physics Laboratory.)

Washington, DC, the light area within the circle, would have been obliterated by the meteor that exploded over Tunguska, Russia. (Photo courtesy of Ball Aerospace and Technologies Corp.)

The circle shows the area of New York and New Jersey that would have been devastated by the Tunguska blast. Millions would have been killed. (Photo courtesy of Ball Aerospace and Technologies Corp.)

Clark R. Chapman, a senior scientist at the Southwest Research Institute, has been at the forefront of those advocating an active, integrated planetary-defense system. (Photo © Carles Ribas [Ediciones EL PAÍS].)

Left: David Morrison, an astrophysicist and the director of the Carl Sagan Center for the Study of Life in the Universe, is a leading specialist in near-Earth objects and began writing *NEO News* in 1984. (Photo from NASA Ames Research Center.)

Right: Former astronaut Russell L. "Rusty" Schweickart was the lunar-module pilot on the Apollo 9 mission and one of the founders of the B612 Foundation. (Photo courtesy of Russell L. Schweickart.)

 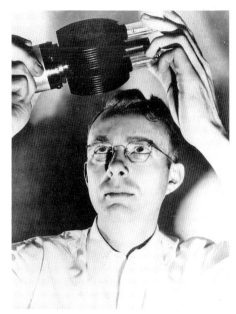

Left: Ed Lu was on two space shuttle missions, spent six months on the International Space Station, and then became one of the founders of the B612 Foundation. (Photo from NASA.)

Right: Luis Alvarez, a Nobel Prize–winning physicist *(shown here)*, his son Walter, and two colleagues were the first to theorize that the huge Chicxulub impact crater off Mexico's Yucatan Peninsula was made in a collision with a near-Earth object that ultimately killed off the dinosaurs. (Photo used by permission of AP Photo.)

Bruce Murray and Carl Sagan *(seated)* and Louis Friedman *(standing, left)* at the time they signed the papers that formally incorporated the Planetary Society. Harry Ashmore *(standing, right)*, a Pulitizer Prize–winning journalist and their advisor, looks on. (Photo courtesy of the Planetary Society.)

Eugene and Carolyn Shoemaker *(pictured)* and David Levy were the first to spot the string of comet fragments that were going to strike Jupiter, and they alerted the world in 1994. It was named Comet Shoemaker-Levy 9 in their honor. (Photo from USGS Astrogeology Science Center.)

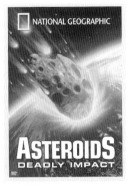

The National Geographic Society produced a special on the asteroid threat. The film included Gene Shoemaker at Meteor Crater in Arizona, which has become a tourist attraction. (Image from *Asteroids: Deadly Impact*, directed by Eitan Weinreich [Washington, DC: National Geographic Video, 2003].)

This diagram of the Sentinel infrared telescope spacecraft shows its large solar-panel array. The cylinder behind it is the telescope. The box near the bottom holds the electronic system, and the white dish at the bottom is a deep-space antenna. (Image courtesy of Ball Aerospace and Technologies Corp.)

sentation, "The Asteroid-Comet Danger and Planetary Defense: A View from Russia," at the opening session of the Schiller Institute conference in Frankfurt am Main, Germany, on April 13–14, 2013. The institute is an international political and economic think tank with headquarters in Germany and in the United States. After the obligatory references to Halley's Comet, Tunguska, and Chelyabinsk, and after noting that Apophis will have a "dangerously close encounter" with Earth in 2029 and 2036 and perhaps cause a "catastrophe on a planetary scale," Benediktov quoted Boris Shustov, the director of the Russian Academy of Science's Institute of Astronomy, as saying that the number of potentially Earth-threatening objects is between two hundred thousand and three hundred thousand—only two percent of which have been identified.

"We need to significantly increase the effectiveness of currently available early warning systems. . . . We need to create a single planetary network for detection and prediction of asteroid and comet hazards." The network should include both already-existing centers, such as the Minor Planet Center, JPL, and the laboratory at the University of Pisa, and new ones, he said. "As for Russia, the work ongoing within individual institutes and research institutions should be systematically organized; a single coordination point has to be set up for data collection and processing. This center should be formed initially as a node of the global (supranational) network."[25]

Then Benediktov startled the audience by quoting Deputy Prime Minister Dmitri Rogozin, his country's representative to NATO, as proposing a civilian-military defensive system that would not only protect the planet from asteroids, comets, "and other space objects," but also from the respective superpowers' intercontinental ballistic missiles; in other words, an international ballistic-missile defense system. "Rogozin stressed that the idea of such a major project under the auspices of the U.N., among other things, gives Russia an opportunity to seize the strategic initiative from the U.S.A. in deploying a global BMD system, including its segment in Europe. It will also make it possible to 'package' a decision on establishing a truly unified and joint European missile defense system into a major civilian project for space exploration in which Russia has its own unique scientific, practical, and industrial role to play. Essentially," Kirill Benediktov said, "Russia and the United States could take on a noble mission to save the planet."[26]

That is what Congress had in mind when it stipulated in the Consolidated Appropriations Act of 2008 that NASA ask the National Research Council, which is the investigative arm of the august National Academy of Sciences, to conduct a study of near-Earth-object surveys and hazard mitigation strategies. That resulted in the creation of the ad hoc Committee to Review Near-Earth Object Surveys and Hazard Mitigation Strategies, which, in turn, formed a Steering Committee, a Survey and Detection Panel, and a Mitigation Panel.[27] The panels met to hear experts—astronomers and other scientists—explain the nature of the situation so the extent of the danger could be determined.

The Survey and Detection Panel met at the National Research Council's headquarters in Washington on November 5–7, 2008, to get the procedure straight. Then it heard testimony there at the end of January 2009, followed by a meeting in Tucson on April 20–22, and a wrap-up session in Santa Fe on July 13–15. The panel's fourteen members were treated to a tour of the Catalina Sky Survey operation while they were in Tucson.

The presentations—scores of them—were appropriately specific and detailed. A representative of Northrop Grumman talked about ASTER, an "Asteroid Structure, Trajectory, and Exploratory Reconnaissance Mission" that could have been sent to size up Apophis. Jon D. Giorgini of JPL and his colleagues described an "Improved Impact Hazard Assessment:

Existing Radar Sites and a New 70-m Southern Hemisphere Radar Installation," while Mark Boslough of the Sandia National Laboratories briefed the panel on "Modeling the Effects of Small NEOs: Low-Altitude Airbursts" (noting that the Tunguska forest was full of dead and rotting trees before that meteor burst clobbered much of what was left standing). And Joseph A. Nuth III of NASA's Goddard Space Flight Center, to take one more example, held forth on "Diogenes A: Diagnostic Observation of the Geology of Near Earth Spectrally-Classified Asteroids."

All the presentations were deftly consolidated into a single volume, replete with excellent illustrations, called *Defending Planet Earth: Near-Earth Object Surveys and Hazard Mitigation Strategies* that was published by the NRC in 2010. It amounted to a briefing manual whose chapters covered risk analysis, survey and detection of NEOs, their characterization and mitigation, challenges of researching the subject, and, inevitably, national and international coordination and collaboration.

The chapter on mitigation showed that the sixteen members of the panel that produced it were well aware of the scenarios that had been worked out by other organizations involved in planetary defense, including the space agency and the B612 Foundation. It therefore suggested four progressively decisive courses of action: (1) civil defense, which would involve evacuating the region around a small impact; (2) the slow push or pull method, which would gradually change the orbit of an NEO so that it misses Earth; (3) the kinetic-impact approach, which would hit the asteroid or comet head-on with so much momentum and energy that it would be abruptly knocked off course; and, if all else fails, (4) using that nuclear warhead to change its orbit. The committee wrote:

> Nuclear explosives constitute a mature technology, with well-characterized outputs. They represent by far the most mass-efficient method of energy transport and should be considered as an option

for NEO mitigation. Nuclear explosives provide the only option for large NEOs (> 500 meters in diameter) when the time to impact is short (years to months), or when other methods have failed and time is running out.[28]

The NRC was also on the same page as the rest of the planetary-defense sector where the need for international cooperation was concerned:

Responding effectively to hazards posed by Near-Earth Objects (NEOs) requires the joint efforts of diverse institutions and individuals. Thus organization plays a key role that is just as important as the technical options. Because NEOs are a global threat, efforts to deal with them may involve international cooperation from the outset. . . . Arrangements at present are largely ad hoc and informal in the United States and abroad, and they involve both government and private entities.[29]

To solve that problem, the committee recommended that "the United States should establish a standing committee, with membership from each of the relevant agencies and departments, to develop a detailed plan for treating all aspects of the threat posed to Earth by Near-Earth Objects." And, the report added, "The United States should take the lead in organizing and empowering a suitable international entity to participate in developing a detailed plan for dealing with the NEO hazard."[30]

That made the course of action compelling and unanimous, but only among the experts in the global community, not among their political leaders.

6

THE DEPARTMENT OF PLANETARY DEFENSE

A specter is haunting NASA (not Europe . . .). It is a victim of its own success, and it is therefore an agency without a mission as profoundly important as the one that got its astronauts to the Moon six times from 1969 to 1972. But there is a new mission, and it is far more important.

NASA's seeds were planted at the imaginatively named First Annual Symposium on Space Travel, which was held at the Hayden Planetarium in New York in October 1951. The event's organizer was Wernher von Braun, the charismatic German aristocrat who masterminded the slave-labor program that made the rocket-propelled V-2 ballistic missiles—the "vengeance weapons"—that rained down on London and elsewhere in the closing stages of the Second World War. He and a little more than a hundred of his colleagues turned themselves and a veritable mountain of blueprints and other technical documents over to the US Army in the closing weeks of the war in an operation first called Overcast and then Paperclip. With the Soviet Union quickly looming as a menace, and the place of the ballistic missile in warfare clearly established, von Braun and six of his top echelon, together with the cache of documents, were spirited to the United States, where they were out of reach of Joseph Stalin's feared Red Army. Six of the "prisoners of peace" (as von Braun called them) were taken to the Aberdeen Proving Ground in Maryland where they were put to work translating, evaluating, and cataloging the documents. Their leader was taken to Fort Bliss, Texas, as the advance man for the rest of the rocket team and eventually to the Army Ordnance Guided Missile Center at the Redstone Arsenal near Huntsville, Alabama, where they designed constantly improved (and ever larger) ballistic missiles. It was at Redstone that von Braun created the gigantic Saturn V rocket that would get six Apollo crews on the Moon.

But von Braun had long since metamorphosed from being a master rocketeer to being a handsome and personable visionary—a space savant—and a publicly recognized celebrity who believed that it was humankind's destiny to explore space and, as the saying of the time went, conquer it. He was to become the face of space for millions of Americans who emerged from the Second World War with a craving to extend their civilization to inhabiting the space around their planet and then extending it to the Moon and to Mars. Walt Disney understood that and capitalized on it by making Tomorrowland one of the four theme parks at Disneyland in California. (One of the others, Frontierland, featured Davy Crockett, who became another American television hero played by Fess Parker, and that had millions of kids wearing coonskin hats, complete with tails.) Typical of Disney's genius, he offered von Braun a consulting job on the project, and the visionary eagerly accepted it. Following that, von Braun took a consultant position on Disney's television show, called *Disneyland*, to work on the "Man in Space" episode, which aired on March 9, 1955, boasting that it was "science factual."

"If we were to start today on an organized, well supported space program, I believe a practical passenger rocket could be built and tested within ten years," von Braun said on camera. "Now here is my design for a four-stage orbital rocket ship . . ." A second segment, "Man and the Moon," which Disney Studios claimed was a "realistic and believable trip to the Moon in a rocket ship," showed von Braun with an excellent prop in his shirt pocket: a slide rule.[1] And he was as knowledgeable and prophetic as he was charismatic, of course, and therefore attracted a coterie of like-minded individuals who had science and engineering backgrounds and who enjoyed writing about it.

They came together at the symposium on space flight, which was held at the Hayden Planetarium in 1951, which was sponsored by *Collier's* magazine. The magazine's editor, Gordon Manning, had come up with the idea. "Within the next 10 or 15 years, the earth can have a new companion in the skies, a man-made satellite which will be man's first foothold in space," von Braun predicted in a reference to what would come to be called the International Space Station (ISS). "Inhabited by human beings, and visible from the ground as a sedately moving star, it will sweep around the earth at an incredible rate of speed in that dark void beyond the atmosphere which is known as 'space.' . . . From this platform, a trip to the moon itself will be just a step, as scientists reckon distance in space." (He was overly optimistic about the time frame, since the first section of the ISS was not carried to orbit until 1998, and it was not manned until late 2000.)[2]

"When man first takes up residence in space, it will be within the spinning hull of a wheel-shaped space station rotating around the earth much as the moon does," science writer Willy Ley told the audience in a presentation on the space station. "Life will be cramped and complicated for space dwellers; they will exist under conditions comparable to those in a modern submarine." Ley was another German-American, a historian of science, and a space advocate. He got the shape of the station wrong but was on the mark about the cramped living conditions. A crater is named after him on the far side of the Moon.[3] Heinz Haber, a German physicist and science writer, delivered a presentation about survival in space that began by enumerating the many mortal dangers, including cosmic and ultra-violent rays and "ultra high-speed projectiles"—the meteors that can easily puncture any protective armor. He concluded his presentation, stating that, while spacefarers would never be completely safe against hazards such as meteors, they could be protected to the point where they "will probably be safer than pedestrians crossing a busy street at rush hour."[4] Joseph Kaplan, a leading geophysicist, described the space environment, while Oscar Schachter, an expert on international law, gave his talk, titled "Who Owns the Universe?" (Everyone and no one, as the master sleuth Sherlock Holmes would have put it.) And Fred L. Whipple, the chairman of the Department of Astronomy at Harvard, ended the session with a presentation called "The Heavens Open," which was appropriately optimistic about humanity's place out there.

Collier's magazine ran eight cover stories that greatly expanded on the meeting, collectively called "Man Will Conquer Space Soon!" as the first installment was named. All of them were illustrated with what was then exotic, futuristic space pictures by Chesley Bonestell, a noted commercial artist. The series ran from 1952 to 1954. The pictures included an astronaut in the "world's first space suit," a spacecraft landing on the Moon (it bore no resemblance to the *Eagle*), and one of another spacecraft approaching Mars. All eight articles were published as three books: *Across the Space Frontier* in 1952, *Conquest of the Moon* in 1953, and *The Exploration of Mars* in 1956.

The meeting at the planetarium and the articles and books that came from it, with their diverse subject matter, were about one overriding fact: humanity's future absolutely and irrevocably involved a permanent presence in space; that space, hostile though it may be, would become an integral extension of its home on Earth. Humanity was destined to "conquer" space, occupy it, and thrive in it indefinitely. It was the critical mass that became the American space program. It was widely accepted by that time that Buck Rogers and Flash Gordon were going to come true; that humankind was poised to take to space for both military and civilian reasons. The operative word for the US presence in space was *control*, meaning not that the

United States would dictate everything that happened there but that nothing could be allowed to happen there that would endanger US national security.

In May 1946, with the guns of the world war barely cooled, the RAND Corp., a California think tank, issued the first in a series of studies ordered by the US Army Air Forces on earth-circling satellites in general and then on reconnaissance "platforms" in particular. The advantages of aerial reconnaissance had been appreciated since Chinese soldiers on kites were used to locate Mongol invaders and follow their movement. And using aircraft to do the same thing—aerial reconnaissance—had played a decisive role in both world wars. As everyone who had ever surveyed a neighborhood from a rooftop knew, the higher the observer, the more could be seen. The fact that a satellite was going to be able to see a great deal more than anyone in an airplane was a given. The first report, *Preliminary Design of an Experimental World-Circling Spaceship*, was completed and submitted in 1948 and was followed by several more specific and detailed studies. It was understood that the unmanned "recce" satellites would be part of an armada of robots that would forever change humanity's relationship with the space around Earth. The reconnaissance and surveillance satellites, together with those that would handle long distance communication, meteorology, navigation, and other important assignments, were to be the unmanned part of the space program. The manned program would be the other part.

The epic meeting at the Hayden Planetarium and the *Collier's* series were extremely important for informing the public on what the budding aerospace community knew was coming. The community included members of the venerable National Advisory Committee for Aeronautics (NACA), which was formed in March 1915 as an emergency organization to promote and coordinate action by the aeroplane industry (the manufacturers of what were then called aeroplanes), academe, and government in fighting the war. But with space operations in the offing, an organization that handled only aviation would obviously be wholly inadequate. One that would oversee all aspects of the civilian space program was obviously going to be needed. It was eventually decided that a single federal entity that was responsible for both air and space operations—they were taken to be a continuum, as the X-15 high-altitude experimental aircraft's forays to the edge of space showed—would be optimally efficient. And that entity would run and coordinate both of them. Advising in that circumstance was clearly a non sequitur. The NACA lived on, but tenuously.

Then there was the Union of Soviet Socialist Republics. While the Soviet military was taken with utmost seriousness, Russia as a whole was the subject of derision and ridicule by people who, hearing that Russians not only claimed to have flown the first airplane in St. Petersburg in the early 1880s but also claimed to have invented baseball, tended to think of them as stoic workers, peasants, and soldiers whose vaunted Red Army turned back Hitler's previously invincible Wehrmacht, but they also thought of Russians as rowdy buffoons; bearded Cossacks in fur hats who danced the kazatsky and whose veins contained equal parts of blood and vodka. So there were "Russian jokes." "What did the Russian people light their homes with before they started using candles?" one apocryphal story asked. Answer: "Electricity." Another had Stalin noticing that there were mice in his study, so he complained to Mikhail Kalinin, the chairman of the Presidium of the Supreme Soviet, who offered this advice: "Why don't you put up a sign saying Collective Farm? Half of the mice will die of hunger and the other half will run away." Alexandr Solzhenitsyn's *The Gulag Archipelago*, a widely read account of his and others' experiences in Stalin's forced-labor camps (published in the West in 1973), confirmed the brutality of the Soviet system and, by implication, the country's barbaric backwardness. The

Russians were a large, resolute, and potentially powerful people who had produced world-class writers and composers, to be sure. Everyone appreciated Tolstoy, Dostoyevsky, and Chekov; Tchaikovsky, Rimsky-Korsakov, and Borodin. But Russians were also thought by their very nature to be nowhere near as consequential in science and technology as the Americans, the British, and the Germans. Dmitry Ivanovsky, who discovered viruses; Ilya Mechnikov, who won a Nobel Prize in Physiology or Medicine as an immune system pioneer; Ivan Pavlov, who won the coveted prize for founding modern physiology; George Gamow, who came up with the big bang theory; Nikolay Basov, who won a Nobel Prize for inventing the laser and the maser; and Vitaly Ginzburg, who developed the Soviet hydrogen bomb, were among hundreds of Russians who showed their country's mettle in the entire array of sciences. (Ginzburg was not awarded a Nobel Prize for coming up with his nation's city-buster . . .) Most Westerners were unaware of Russian scientific and medical advances, however, and believed that the Russians were definitely underachievers in those areas.

That abruptly changed on October 4, 1957, when a huge R-7 rocket lifted off the concrete launchpad at the new launch facility at the Baikonur Cosmodrome on the Kazakh Steppe and tossed an 83.6-kilogram, polished metal sphere called *Sputnik* into Earth orbit. It was part of the Soviet Union's previously announced contribution to the International Geophysical Year (IGY), which was the first major study of Earth by its leading scientists of several nationalities. *Sputnik* carried a radio transmitter, three large batteries, and two sets of antennas so it could broadcast its position to everyone with a radio. Its purpose was to send down information on the density of the extreme upper atmosphere and the ionosphere. More important, it was politically as well as scientifically important, since it required a substantial scientific and technological capability that surprised and impressed many Americans.

C. Turner Catledge, the managing editor of the *New York Times*, and his colleagues, in common with their opposite number at the *Washington Post* and other good-morning newspapers, knew a very big story when they saw one and therefore ran a headline across the top of page one that proclaimed:

SOVIET FIRES EARTH SATELLITE INTO SPACE;
IT IS CIRCLING THE GLOBE AT 18,000 M.P.H.;
SPHERE TRACKED IN 4 CROSSINGS OVER U.S.

Most everyone who was paying attention was impressed. Rear Admiral Rawson Bennett was a notable exception, or so he claimed in public. Bennett was the chief of naval operations and was therefore the nominal head of the Vanguard Program, which was supposed to orbit an American satellite in the IGY. He dismissed *Sputnik* as "a hunk of iron almost anybody could launch."[5] That was the kind of disingenuous statement that quickly gave rise to what was called the *Sputnik* cocktail: equal parts vodka and sour grapes. As the nominal head of the Vanguard Program, Bennett presided over eight launch failures and only three successes. From 1957 to 1959 evening television-news audiences grew accustomed to seeing the pencil-shaped rockets either blow up on the launchpad without moving or rise a few feet and then come slowly, agonizingly, down, their motors firing to no avail, and disappear in a growing black-and-white fireball.

The *Sputnik* cocktail was savored by MIT undergraduates, too. The institute's humor magazine, *Voo Doo*, ran a cartoon in the November 1957 issue showing a dopey-looking Cossack huddled inside a cutaway *Sputnik*, microphone in hand, saying ". . . beep—beep—beep—beep—beep—beep—beep . . ."

Sputnik's taking to space was broken in a routine Radio Moscow announcement three hours after its launch, when its achieving orbit was a certainty. It flew over the United States twice, racing smoothly past a starry background, before anyone realized it was there. But then, when it sank in that a Soviet spacecraft was flying imperviously over their country, there was a storm of self-flagellation and recrimination, much of it directed at President Dwight D. Eisenhower, who was widely perceived as being an amiable man who was not intellectually up to serious statecraft; who was more interested in golf than in national security. That was manifestly wrong, as his advocacy of the Open Skies program and strong support of the nation's ballistic-missile program showed. His authorization of U-2 flights over the Soviet Union that began in 1956 and his approval of the *Corona* satellite reconnaissance program that immediately followed them showed that he was well aware of the Soviet threat and was acting aggressively to counter it. He gamely answered his critics by insisting—correctly—that the United States did not trail the Reds in science and technology. It was not *Sputnik* that troubled Ike, the National Security Council, and the Joint Chiefs of Staff, but the R-7 that sent it to orbit. They knew that a rocket that could send an 83.6-kilogram satellite all the way to space could also send a much heavier nuclear warhead to the United States over the Arctic at a much lower altitude.

On October 1, 1958, almost a year to the day after *Sputnik* took to the sky and caused a lingering trauma, the NACA was transformed into the National Aeronautics and Space Administration (NASA) as a tacit acknowledgment that the Space Age had indeed dawned and that a permanent organization on the order of the Department of Defense and other federal departments was needed to run all US civilian space activities, manned and unmanned. (The armed services and the CIA would start their own, often competitive, space programs.) And in recognition of the air-space continuum, NASA was mandated to be responsible for both sectors. Less than three years later, the Russkies struck again.

On April 12, 1961, an R-7 roared out of Baikonur, carrying Yuri Gagarin in *Vostok 1* on a complete orbit around Earth, making him the first mortal to reach space. *"Poyekhali!"* ("Here we go!") said the exhilarated cosmonaut as the modified R-7 lifted slowly off the launchpad and began its climb to orbit at a little after nine o'clock on the morning, as Sergei Pavlovich Korolev, the space program's ingenious chief designer (its Wernher von Braun, and the man who conceived *Sputnik*), and Valentin Glushko and Mstislav Keldysh (two colleagues), watched through periscopes in a nearby bunker. The twenty-seven-year-old test pilot made one complete orbit of the planet before climbing out of the capsule and parachuting separately to earth. To their other firsts, the Russians now added getting the first man to space.

On April 13, under a "hed" that spanned four of the newspaper's eight columns—"RUSSIAN ORBITED THE EARTH ONCE, OBSERVING IT THROUGH PORTHOLES; SPACE FLIGHT LASTED 108 MINUTES"—the *New York Times* ran a photo of jubilant youngsters outside the Moscow Planetarium and another of their hero with the barest trace of a smile, like a cosmic Mona Lisa in a leather flight cap (he had worn a helmet to space).

The "newspaper of record" ran a hastily assembled spate of sidebars that described the flight with maps and diagrams, provided background on the preparations for the mission, and even included a transcript of some of the hero's radio chatter with his controllers at the space facility at Kaliningrad. It also carried congratulations from President John F. Kennedy. NASA's chief designer, Wernher von Braun, also offered congratulations. The *Times* ran man-in-the-news profiles that were supposed to provide depth—the human "angle"—about newsmakers. Gagarin's showed a photograph of him beside his wife, Valentina, who was reading to their two-year-old daughter. In a reference that was to too good to ignore, the writer of the piece noted that "Gagarin" derived from "wild duck." And there was a cartoon from the *Baltimore Sun* that showed Premier

Nikita Khrushchev holding a red star in space with one hand and, in the other, a shoe, with which he was banging a likeness of Kennedy over the head.[6] (The shoe banging was a reference to Khrushchev's banging his shoe on his desk at a UN General Assembly meeting in the autumn of 1960 to protest the Philippine delegate's public reference to Eastern Europeans and others as having been deprived of their civil and political rights and "swallowed up" by the Soviet Union. It was widely used to portray Khrushchev as an obstreperous Slavic lout who lacked the poise that was necessary to be a statesman.)

"Bourgeois statesmen used to poke fun at us, saying that we Russians were running around in bark sandals and lapping up cabbage soup with those sandals," Khrushchev told a Polish audience two years after Gagarin's flight. "They used to make fun of our culture, the culture of people considered, so to say, to be the last among the civilized Western countries. Then suddenly, you understand, those who they thought lapped up the cabbage soup with bark sandals got into outer space earlier than the so-called civilized ones." [7]

Gagarin's feat, like *Sputnik*, made page-one headlines around the world. The hero was duly photographed in uniform with Nikita Khrushchev, Leonid Brezhnev, and other dignitaries on Lenin's Tomb two days after his flight and separately with an obviously respectful Korolev. The Motherland was so proud of Korolev that it turned his home into a museum, and so proud was it of the other designers and cosmonauts—and the fact that their nation had started the Space Age—that the Memorial Museum of Cosmonautics was opened at the intersection of Mir Avenue and Academician S. P. Korolev Street in Moscow on April 10, 1981, two days short of the twentieth anniversary of Gagarin's flight. *Mir*, which means "peace," was the name of the Soviet space station. The large museum was filled with artifacts from that glorious time, including a *Sputnik*, complete with antennas; large models of the giant rockets that launched the manned and unmanned spacecraft; the *Vostok* spacecraft in which Gagarin made his epic flight; Veneras and other deep-space probes that represented the missions to Venus; spacesuits and other clothing; and art that depicted *Sputnik*, Gagarin in an orange flight suit surrounded by fluttering white doves, and cosmonaut Aleksei Leonov floating over the Black Sea; commemorative coins and pennants; tubes of space food; and posters that paid clear tribute to the space program. A typical poster by cosmonaut-artist Alexei Sokolov, titled *Glory to the Conquerors of Space*, showed Alexei Leonov, the first man to float outside a spacecraft (and the eventual vice president of Alpha Bank) heading for the Mir space station in a *Soyuz* capsule.

On August 6, 1961, Gherman Titov flew in *Vostok 2*, accomplishing the first manned mission that lasted a full day. Almost exactly a year later, Andriyan Nikolayev and Pavel Popovich became the first to fly in formation in *Vostoks 3* and *4*. On June 14, 1963, cosmonaut Valery Bykovsky made the longest solo orbital flight in *Vostok 5*, and, two days later, Valentina Tereshkova became the first woman and the first civilian to get to space when she was carried there in *Vostok 6*. She not only orbited the world forty-eight times, which was more than the total for Alan Shepard, Gus Grissom, John Glenn, and Scott Carpenter, America's first four Mercury astronauts, combined, but she and Bykovsky were connected by radio and flew in a virtual formation, which was planned to be as impressive as it was. Airplanes flew in formation and now, thanks to Soviet innovative leadership, so did spacecraft. And to make matters slightly worse for the Americans, the fact that Tereshkova was a woman subtly suggested—however incorrectly—that the social as well as economic equality that Marxism-Leninism promised was obviously true in "the people's paradise." Sally Ride, America's first woman astronaut—and the holder of a doctorate in physics from Stanford University—got to space on the shuttle *Challenger* on June 18, 1983, almost exactly two decades later.

First quickly became the buzzword in the United States and elsewhere in the West as Korolev and his compatriots racked up one after another. The unstated but widely accepted implication was that a nation that was first in performing all of those feats had an energetic and robust space program with skilled personnel and excellent facilities. R-7s and other giant lifters were not fired from slingshots. The infrastructure, let alone the science and engineering, that was required to get people to space was formidable and reflected a basic strength that necessarily had to be extensive. Being able to stage such spectacular performances, in other words, was the sure mark of strength that extended to relations with the rest of the world. It made the difference between being a mere power and being a superpower.

John F. Kennedy, who was so notoriously competitive that he even hated losing at touch football, knew that. The consecutive, daring Soviet achievements got the rapt attention of the American news media and briefly seemed to cast doubt on his country's leading the world in science and technology. It seemed to betray Edison, Morse, Ford, Lindbergh, Perry, Byrd, and the other inventors, adventurers, and explorers. It rankled. Space, as the Caltech geologist and space scientist Bruce Murray has said, is a reflection of Earth.[8] The prospect of the "greatest nation on Earth" (as the United States called itself) being repeatedly beaten in space and therefore seeming to be a second-rate power, at least in that realm, was not acceptable to JFK. Nor were things much better on terra firma. The civil-rights situation in the south had turned explosive; the Communist Pathet Lao were dangerously close to toppling the pro-American government in Laos; the Communist-led Vietcong, aided by North Vietnam, was waging a war against the pro–West South Vietnamese government; and the CIA-backed invasion of Cuba to remove Fidel Castro from power turned into a fiasco when the invading force of Cuban exiles, lacking the air cover they expected, was almost massacred at the Bay of Pigs. Then space began to turn red, or so it seemed in the White House, with an impressionable Third World watching. The Underdeveloped World, as it had been called until recently, was taken—incorrectly—to be up for grabs in terms of Eastern or Western "influence." Its leaders tended to play both sides against each other while steadfastly maintaining their independence.

"The President was more convinced than any of his advisers that a second-rate, second-place space effort was inconsistent with the country's security, with its role as world leader and with the New Frontier spirit of discovery," Theodore C. Sorensen, Kennedy's special counsel, recalled years later.[9]

On the basis of advice Kennedy got from his inner circle, including Sorensen, he therefore made what he later called one of the most important decisions of his presidency: "to shift our efforts in space from low to high gear."

On May 25, 1961, in a speech to Congress that addressed "urgent national needs," Kennedy mentioned several dangers that faced the United States, including Communist subversion. He then used the *Sputnik* and Gagarin flights, and their impact on "the minds of men everywhere," to call for the United States to land a man on the Moon and return him safely to Earth before the end of the decade. "While we cannot guarantee that we shall one day be first," he warned, "we can guarantee that any failure to make this effort will find us last." He repeated the plan in an address at Rice University in Texas on September 12, 1962, and that was the one that got the news media's attention and put him all over page one.

> We choose to go to the Moon. We choose to go to the Moon in this decade and do the other things, not because they are easy, but because they are hard, because that goal will serve to organize and measure the best of our energies and skills, because that challenge is one that we are willing to accept, one we are unwilling to postpone, and one which we intend to win, and the others, too. It is for these reasons that I regard the decision last year to shift our efforts in space from low to high gear as among the most important deci-

sions that will be made during my incumbency in the office of the Presidency.[10]

Announcing only that Americans were going to go to the Moon would have been very good strategy. It would have been a fine foreign-policy ploy, something of a morale booster for the American people, and an encouraging signal to the space agency and the science and industrial sectors. But putting a time limit on it—within the decade, meaning within eight years—was brilliant. The finite time period was a virtual guarantee that, unlike so many promising programs in Washington that evaporate after they are announced, landing men on the Moon was really going to happen because an actual, stated, on-the-record deadline made it definite. The world had the president's word on it, and no president would volitionally lie, particularly about a subject of such importance whose failure was guaranteed to diminish his image in the history books. Furthermore, deadlines are inherently dramatic because they create the pressure that comes with competition, and there is a penalty of some sort when they are not met. That is why sports events are played in finite periods of time that are extended only if the competitors are tied. "Third and goal with less than a minute on the clock" generally gets football fans on their feet. A tie score that would extend the game to the next day would not.

With that announcement, John M. Logsdon, a leading space expert and author of *The Decision to Go to the Moon: Project Apollo and the National Interest,* has rightly explained, Kennedy not only set a single, overarching goal for the space program, but he also fundamentally changed the nature of the program itself. "He challenged the assertion that a 'single civil-military program . . . is unattainable' by approving the initial plan for just such a program, aimed at establishing American preeminence in every aspect of space activity, civilian and military, scientific and commercial, prestige-oriented and unspectacular."

He thereby abruptly and dramatically reversed Eisenhower's decentralized and lackluster space program.[11] JFK, in other words, finally got a fractured and unfocused operation pointing in a single direction with a dramatic goal that would require heroism and dignify the whole human race (the Russians, the Chinese, and other Cold War foes, plus the Third World).

Kennedy mandated NASA to plan and carry out no less than the single greatest and most dramatic feat of exploration in human history: to send men to another world. The space agency managed to do that not once but seven times, with six crews landing on the Moon in a program that was adroitly christened Apollo, after the god of light and the Sun in Greek mythology, an omnipotent oracle who bestowed truth and culture on the world. The Russians were decisively routed, and bragging rights were decisively and dramatically won by the land of the free and the home of the brave.

All of the free world cheered when Neil Armstrong and Buzz Aldrin alighted on the Sea of Tranquility while Michael Collins cruised overhead in the *Apollo 11* command module. The landing captured the imagination of the world, with untold millions watching it on television, listening to it on radios, and reading about it in their newspapers (including *Pravda* and *Izvestia* in Russia, and *Jen-Min Jih-Pao* in China). Their fellow men had landed on another world. That was taken to mean that, given the resolve, anything was possible. It was literally the high point in human history—a transcendental moment when humanity expanded its domain as never before.

It was also a masterpiece of subtle public relations. "That's one small step for man, one giant leap for mankind," Armstrong famously proclaimed as he set foot on the lunar surface to begin a romp that lasted almost three hours. That meant the landing was on behalf of all humanity; that Americans had ventured to another world as representatives of all humankind, a distinctly noble gesture. But the flag that Aldrin planted in the lunar soil as he was being photographed in color was not the United Nation's. It was the stars and stripes; Old Glory.

The newspaper of record (as the *Times* called itself) devoted all of page one to the story, as did every other reputable paper in the country:

MEN WALK ON MOON
ASTRONAUTS LAND ON PLAIN;
COLLECT ROCKS, PLANT FLAG

"Men have landed and walked on the moon," John Noble Wilford, the reporter who covered the space program for the *Times*, wrote with eloquent simplicity at the Johnson Space Center in Houston, where Mission Control controlled the mission.

> Two Americans, astronauts of Apollo 11, steered their fragile four legged lunar module safely and smoothly to the historic landing yesterday at 4:17:40 P.M., Eastern daylight time.
>
> Neil A. Armstrong, the 38-year-old commander, radioed to earth and the mission control room here:
>
> "Houston, Tranquility Base here. The Eagle has landed."[12]

The space establishment was of course ecstatic, and so, for the most part, were the Pentagon, the news media, the industrial sector, highly educated professionals, Joe Six Pack, and ordinary Americans everywhere.

The intellectual establishment had three perspectives: that there were higher priorities on the home planet, such as disease and poverty; that the Moon ought to be untouched by humans and left in its pristine condition; and that landing on it would not only be scientifically important but would be the greatest adventure of all time, and one that helped unify the world.

Arnold J. Toynbee, the venerable British historian, thought that landing on the Moon symbolized a large gap between technology and morals. "In a sense," he said, "going to the Moon is like building the pyramids or Louis XIV's palace at Versailles. It's rather scandalous, when human beings are going short of necessities, to do this. If we're clever enough to reach the Moon, don't we feel rather foolish in our mismanagement of human affairs?" Mark Van Doren, the Columbia University poet and professor of English, thought that the Moon was majestic and a symbol of nature and the universe because it was unsullied by humans and ought to remain that way. "I wish we would leave the Moon alone. I have great respect for the Moon. The arrogance of men landing on the Moon is, to me, very shocking and painful," he told a journalist.[13] Reinhold Niebuhr, a leading theologian, agreed.

Predictably, scientists strongly disagreed. "The Roman Empire decayed when it ceased to be progressive in this kind of sense, and there are other examples," Sir Bernard Lovell, the director of Britain's Jodrell Bank Observatory, contended. "To a certain extent, you see the beginnings of it in the United Kingdom today, but fortunately not in the United States and certainly not in the Soviet Union."

Margaret Mead, an animated anthropologist at the American Museum of Natural History in New York, heartily agreed. "People have always said that it would be better to stay at home and till your own cabbage patch. I think that if people don't follow the potentialities of movement and change, they're likely to wither and die," she said, adding that we would "hate ourselves" if we did not go there.[14]

And Isaac Asimov, who had a degree in biochemistry that he applied to science fiction and who was then publishing his hundredth-or-so book, saw Earthlings going to the Moon as a unifying factor. "Once we reach the Moon," he said, "I think we will have made our point and should stop fooling around. The trip to Mars will be too expensive for either the United States or the Soviet Union to do alone. This is an age of global problems. By combining for the conquest of space, we can cooperate where it bothers our prejudices the least because none of us has a vested interest in space. During the International Geophysical Year, for example, everyone agreed on the manner in which they would investigate Antarctica. It was an empty land which belonged to nobody and they could agree on it without loss of face. Similarly, we can agree on space."[15] The exploration of the Moon, then, should be an international operation that would help unify nations—and, by implication, promote peace—and would lead the way to Mars.

The *Times* enthusiastically supported the Apollo program and a presence in space in general. But, to its chagrin, that had not always been the case. On January 13, 1920, "Topics of the Times" ran a short, smug editorial-page feature that basically said that space travel was impossible. It dismissed the notion that a rocket could function in a vacuum, where there was nothing to push against, and ridiculed Robert H. Goddard, the American rocket pioneer, for having the temerity to believe otherwise. "That Professor Goddard, with his 'chair' in Clark College and the countenancing of the Smithsonian Institution, does not know the relation of action to reaction, and of the need to have better than a vacuum against which to react—to say that would be absurd. Of course he only seems to lack the knowledge ladled out daily in high schools." The newspaper's editorial board saw fit to run a correction on July 17, 1969, almost a half century later, as Armstrong, Aldrin, and Collins headed for the Moon, that was cleverly self-mocking. Under a headline that read "A Correction," the story recounted the mistake and concluded with dry humor that "Further investigation and experimentation have confirmed the findings of Newton in the 17th century and it is now definitely established that a rocket can function in a vacuum as in the atmosphere. The Times regrets the error."[16]

That first Moon landing was NASA's finest moment. Armstrong, Aldrin, and Collins returned from the Sea of Tranquility to a ticker-tape parade in New York and parades in Chicago and Los Angeles. They were awarded the Presidential Medal of Freedom by President Richard M. Nixon and Vice President Spiro T. Agnew and had a forty-five-day Giant Leap tour of twenty-five countries that included an audience with Queen Elizabeth II. All of America cheered them, and, implicitly, the space agency that got them to the Moon and back. Many proclaimed almost deliriously that they felt privileged to live at a time when their race first went to another world; it was a moment unique in all of history and they were deeply happy to witness it. But the exhilaration and enthusiasm quickly lessened because of a natural falling off of interest in repeat performances, and also because of détente with the Evil Empire, as President Reagan would call the USSR. And the counterculture was taking hold, and with it, a restructuring of many Americans' priorities. The nation's collective mind was convulsing over the war in Vietnam; the civil-rights movement was fighting for racial equality and an end to bigotry (in the north as well as in the south); and self-proclaimed public-interest groups were demanding a reversal of urban decay, improved public education, and an end to crippling inflation that was severely hurting the poor and the middle class. Campus unrest, mostly because of Vietnam, but also because of the racial situation, was rampant.

No wonder public interest in the Apollo missions had fallen off substantially by the time Charles "Pete" Conrad Jr. and Alan L. Bean spent more than a day on the Sea of Storms while Richard F. Gordon Jr. orbited in the command module in November 1969 on the Apollo 12 mission. And if the distractions were not enough, an oxygen tank exploded in Apollo 13's service module on April 13, 1970, two days after launch, forcing it to return home without landing on the Moon. The accident brought to mind the Apollo 1 fire at Cape Canaveral on January 27, 1967, that killed Virgil I. "Gus" Grissom, Roger Chaffee, and Edward H. White. The Apollo 14, 15, 16, and 17 missions were duly flown, with the last landing in the Taurus-Littrow valley and spending a record three days there, including a record twenty-two hours on extravehicular activity, as being outside a spacecraft in space or on *luna firma* is called, collecting a record 110.5 kilograms of rocks and other material and leaving scientific instruments. Eugene Cernan, Ronald Evans, and Harrison Schmitt (a geologist and the only scientist to land on the Moon) set a record for the longest time on the Moon before they came home on December 7, 1972. By then, the planned Apollo 18, 19, and 20 missions had been scrubbed for a lack of both funding and public interest.

Alan Bean told John Noble Wilford, the *New York Times*'s space reporter, that the Apollo astronauts had taken it for granted that the program they started would continue with the construction of a lunar base and space stations (plural) as part of humanity's logical expansion to space for a permanent presence there. "At that time in our culture's history, we were doing the most that was possible to be done. We naively assumed that's what would continue, but it didn't," a disappointed Bean reflected. "It's the normal thing for a culture, in history, that we respond to emergencies."[17]

They are looming. The litany of dangers, from high-velocity boulders peppering the neighborhood to resource depletion, to continuing terrorism, to global warming and the multiple problems it is causing, to overpopulation. Yet humanity is caught in a dangerous predicament. Unlike the other creatures on this planet, humans—at least some of them—have the intellectual capacity to understand the precariousness of the situation. But there is no inclination to respond to it with a long-term plan because, like the other creatures, humans are fundamentally—perhaps because of their evolution—incapable of projecting threats to the distant future and coming up with ways to reduce or avert them. It is the ultimate chess game, and we are playing it like wood-pushers.

E. O. Wilson, the naturalist, has his own theory, which he shared in a speech he gave to the members of the Foundation for the Future in August 2002, when he was presented with an award. "The human brain evidently evolved to commit itself emotionally only to a small piece of geography, a limiting band of kinsmen, and two or three generations into the future. We are innately inclined to ignore any distant possibility not yet requiring examination, however promising, or menacing." He explained it in Darwinian terms. "For hundreds of millennia, those who worked for short term gains in a small circle of relatives and friends lived longer and left more offspring, even when—and this is the important part—their collective striving put their descendants at risk."[18]

Apollo's end left NASA in a quandary. Although it ran many routine satellite operations that provided weather, communication, and navigation information that had become indispensable to a modern society, Apollo was by far its greatest, most ennobling achievement, and it was now history. Next in order of prestige and public awareness was the exploration of the Solar System, which had started with the Ranger and Surveyor projects scouting the Moon in preparation for Apollo, and then it extended to the Mariner missions to Mercury, Venus, and Mars; the Pioneer missions to Jupiter and Saturn and to investigate solar phenomena; the landing of two Viking spacecraft on Mars; the Voyagers' sensational Grand Tour; Magellan's mapping of Venus; and Galileo's intense reexamination of the King of the Planets. The "take" from the science missions was phenomenal and ranks with the exploration of Earth itself in the eighteenth and nineteenth centuries. But the accumulation of all the information, together with budget cuts because of priorities at home, eventually brought Solar System exploration to near closure as well. The same for the International Space Station, which, because of the retirement of the shuttles, Americans can reach only by paying Russia—which "lost" the space race (whatever that was)—to transport them in *Soyuz* spacecraft. And with the days of Chuck Yeager and his X-1, followed by the extraordinary X-15, long gone, aeronautical testing is routinely going on but is negligible. That leaves NASA without a major mission. But there is one: planetary defense.

The agency without a major mission employs some eighteen thousand people in its Washington headquarters and in ten field centers spread around the country, most notably at the John F. Kennedy Space Center on Florida's east coast, which used to be Cape Canaveral, and which has launched all civilian missions, human and robotic, since the space program began. The Lyndon B. Johnson Space Center in Houston is the human-spaceflight center. In keeping with the state's outsized reputation, it is a complex that consists of one hundred buildings on 1,620 acres and trains US astronauts and other nations' spacefarers. The George C. Marshall Space Flight Center at the Redstone Arsenal in Huntsville, Alabama, where von Braun presided and where the *Saturn V* was therefore built, is where the shuttle was designed and produced, as well as the International Space Station. The John H. Glenn Research Center in Cleveland does what its name says: it researches advanced technology for both air-breathing and rocket systems. JPL, the Jet Propulsion Laboratory, in Pasadena, ran the Voyager and other Solar System–exploration missions, including those Ranger and Surveyor landings that scouted the lunar surface in preparation for Armstrong, Aldrin, Collins, and the eighteen astronauts who followed them. And the Ames Research Center at Moffett Field in California, to take one more, did fundamental research on robotic lunar exploration and now does it on spaceflight, information technology, and, what is most relevant to planetary defense, on the objects that prowl around the neighborhood. NASA also runs the Deep Space Network (DSN), which operates large, movable, parabolic radar antennas in Goldstone, California; at a site west of Madrid; and at Canberra, Australia, that watch the sky in all directions. (The network is described in the next chapter: "The Ultimate Strategic Defense Initiative.")

With the overriding demands of the manned program effectively over and Solar System exploration now marginal, the space agency's considerable human and physical resources ought to be engaged in protecting Earth from near-Earth objects (NEOs) as its highly focused, central mission; its chief raison d'être. That is not to say that other programs should be abandoned. We must go on exploring, not only to expand our presence in the universe, but also to increase the knowledge that defines us. And obviously NASA's routine but imperative services must continue. But Annalee Newitz, a science writer and the author of *Scatter, Adapt, and Remember: How Humans Will Survive a Mass Extinction*, got it right when she wrote that "today, we have evidence that confirms environmental changes like these [the widespread devastation caused by a major impact] can be blamed directly or indirectly for most mass extinctions that have scourged the Earth. And that's why our space program isn't just something educational we're doing to learn more about the universe. It's vital to our survival as a species, because the Earth isn't going to be a safe place for us in the long term."[19]

Planetary defense is so important that it warrants being run by a separate, specialized organization within NASA, perhaps called the Department of Planetary Defense, whose staff would come not only from NASA but also from the four military services. Since the threat is worldwide, the DPD would work closely with those who are doing similar work in foreign space organizations and with the United Nations and NEOShield, an international planetary-defense group that was started in 2012 specifically to foster coordinated collective defense of the planet.

It is extremely important that the Department of Planetary Defense have a budget that is adequate to maintaining a serious program and that the necessarily long-term strategy it would adopt not be subject to the vagaries of congressional politics, particularly as they would affect that budget. In that regard, the DPD would cooperate with foreign space agencies and NEOShield to the fullest extent possible and be flexible enough to adapt to useful suggestions, but as is the case with other US federal institutions such as the Defense Department, it would not be bound to obey their suggestions. The Defense Department is an American member of NATO, but it does not take orders from it.

The Department of Planetary Defense would also maintain liaison with the planetary-defense groups, such as the Planetary Defense Foundation and the B612 Foundation, as well as the national and international scientific establishments, because ideas are generated in those environments, often synergistically. The National Research Council's work in survey and mitigation strategy could benefit the DPD, for example, and so could a defensive system that consists of Rusty Schweickart and Ed Lu's three-phase planetary defense—locate and identify potential impactors, try to nudge them off course, and destroy them a long way out if nudging does not work—plus a last-ditch defense that uses specially adapted technology from the ill-fated (and ill-conceived) Strategic Defense Initiative, or Star Wars, scheme that was considered during the Reagan administration.

The weapons the Department of Planetary Defense and NEOShield could use to obliterate approaching NEOs—and that is precisely what they would be and should be called: *weapons*—are described in chapter 7. But weapons must be the last resort because, again, blowing up an asteroid that is relatively close and approaching at high velocity would turn it into fragments that could inflect terrible damage independently. The goal is to spot attackers so far ahead of impact—a decade or more is preferable—that they can be gently nudged off course long before they have to be stopped with weapons because their threat is becoming immediate. That capability will obviously require an excellent long-range surveillance system that sees in all directions.

THE ULTIMATE STRATEGIC DEFENSE INITIATIVE

ntelligence is defined as the ability to learn and understand new situations, some of them extremely complicated, and to apply that knowledge to manipulate the environment to one's advantage. That especially applies to situations that are dangerous or potentially dangerous.

Avoiding danger is taken to be so imperative by most of the nations in this world that they have created special organizations whose purpose is to find anything that constitutes a danger to the security of the state and then reduce or eliminate it. Having effective intelligence—knowing precisely what is going on beyond the nation-state's borders—is therefore an integral part of the operation of every government in the world. Intelligence professionals hate surprises because they can cause terrible damage and devastating defeat. The rule is therefore to know as much as possible about what is happening in one's environment (or one's space, as it is now appropriately called) all the time. It is broadly known as situational awareness, and it began in the caves, continued to the city-states, then to nations, and, as the world coalesced in the twentieth century, to this planet as a whole. The landings on the Moon—extending humanity to another world in the great tradition of exploration—was ennobling. The imagery taken of Earth showed a beautiful solitary planet, an oasis of life, slowly rotating in black space and, seen against the airless void that surrounds it, made it clear that its creatures are entirely dependent on it for their survival and that it is fragile, vulnerable, and perishable. Some of the rocks that speed by constitute a threat, as the craters both on Earth and on the body on which the astronauts stood taking pictures, clearly showed.

The US Space Surveillance Network was started in 1957 immediately after *Sputnik* went into orbit. It is a critical part of US Space Command's global presence and is responsible for detecting, identifying, tracking, and cataloging manmade objects orbiting Earth, including both active and inactive satellites, spent rockets, and debris that range in size from small chunks of broken satellites to entire rocket upper stages, to the International Space Station. The network scrutinizes space for reasons of national security, and that obviously means knowing what foreign spacecraft, civilian as well as military, are out there and what they are doing that could potentially threaten the national security of the United States.

But security in its broadest sense means freedom from danger—safety—and it has a far larger dimension than the military "arena." As is the case with individuals, safety means that the nation will not come to harm and that its well-being is assured to the greatest extent possible. Safety for the individual not only means that he or she will not be manhandled, mugged, or murdered, but also that he or she will not be killed by a lamp that is tossed out of a sixth-floor window or by an automobile that runs a red light through a crosswalk; in other words, be the victim of a fatal accident. It is the same with nations, where security not only requires that a country is safe from foreign military attack (and now terrorism), but that it also is safe from major attacks by nature. Defenses have therefore been put in place to increase the safety margin from natural disasters, which is why hurricanes and typhoons are tracked by imaging satellites and Caribbean hurricanes are penetrated by US Air Force Hurricane Hunters crews that fly into them to measure their intensity; the Richter scale is used to calculate the energy released during an earthquake so scientists can gain information about them and are able to better predict their force; and icebergs and ice floes are monitored to avoid a repetition of the *Titanic* tragedy.

And national and international security, to carry the analogy further, have dimensions that go well beyond the sanctity of boundaries and air space, freedom of the seas, and limits on nuclear-weapons stockpiles. They also include safety from sudden, devastating attacks by nature. The most potentially dangerous of those attacks, as Gene Shoemaker said, could (and have) come from large (and not-so-large) objects flying near Earth at very high velocity. After prevention of a nuclear war, staying safe from asteroids and comets that could cause terrible death and destruction, if not humankind's extinction, has therefore been one of the world's most important priorities. And it did not take a Stephen Hawking, a Freeman Dyson, or a Richard Feynman to know that potential impactors could not be prevented from slamming into Earth unless their whereabouts were precisely known years before impact and their trajectories accurately established. That is why Congress ordered NASA to do the Spaceguard Survey in 1998 and have it completed by 2008.

But by then, the astronomers, astrophysicists, and others interested in the impact threat had already long been addressing it. They held an International Near-Earth Object Detection Workshop in Vail, Colorado, that began in May 1991 and ended in January 1992, conducted under NASA's auspices. The participants decided that wayward asteroids and comets are indeed a significant hazard to life and property and accepted the mantra that while the possibility of being hit by a large one is extremely small, the consequences of such a collision "are so catastrophic that it is prudent to assess the nature of the threat and prepare to deal with it." The first step in preventing a catastrophe, the astronomers and others rightly decided, requires a comprehensive search for Earth-crossers and a detailed analysis of their orbits. The workshop therefore produced a report that contained its deliberations on creating a program that would dramatically increase the detection rate of Earth-crossers.

Detecting "intruders" is one thing. What to do when it is determined that they are bearing down on Earth is another, and, appropriately, that subject came up at an interception workshop that was held at the Los Alamos National Laboratory in early 1992. Having determined what the threat was, the next logical step was to try to come up with ways to mitigate it. That meant devising ways to nudge potential impactors off course a decade, and ideally a couple of decades, ahead of time. (Everyone agreed that the Bruce Willis Defense, as shown in the film *Armageddon*, in which a comet is nuked a week or so before the collision, is great theater but abominable planetary defense since it would merely turn a cannonball into grapeshot, each chunk of which could destroy a city.) Edward "The Father of the H-Bomb" Teller and Lowell Wood nevertheless gave presentations that called for blasting them with nuclear weapons. That led to what many at the meeting called "the giggle factor," according to Clark R. Chapman. Wood was an astrophysicist at the Lawrence Livermore National Laboratory in New Mexico, a disciple of Teller's, and the inventor of the nuclear-pumped x-ray laser, which was supposed to destroy Soviet ballistic missiles by hitting them with laser beams coming from nuclear explosions in the stratosphere as part of President Ronald W. Reagan's Strategic Defense Initiative (which is described further on).

"Many of the Department of Energy researchers who had been recently introduced to the question of how to deflect asteroids, were provided with a technically flawed analysis by Wood et al., arguing that there is great danger from very small asteroids that impact frequently; in fact, they burn up harmlessly in the Earth's atmosphere, never posing a danger to anyone on the ground," Chapman has written. He continued:

Nevertheless, because of the Wood analysis, many of the analyses readied for presentation at Los Alamos concerned application of Strategic Defense Initiative ("Star Wars") technology originally intended to shoot down enemy missiles, which aren't very different from Wood's cosmic bullets. Astronomers in the audience, including Morrison and myself, tried to explain that the small objects posed no significant hazard, and that it was the large (> 1.5 km diameter) objects that were of concern. Indeed, schemes *were* discussed at the Detection Workshop that would mitigate impact of the larger ones, including stand-off detonation of a neutron bomb, but also including outlandish "blue sky" ideas involving anti-matter and other notions more appropriate to science fiction novels.

Chapman so disagreed with using a Star Wars defense against near-Earth objects (NEOs) that he exercised the ancient prerogative of the coauthors of scholarly papers who disagree with their colleagues' conclusions. He asked that his name be omitted from the report, and, as custom required, his request was honored.[1]

The planetary-defense community, which had long understood the NEO threat and believed that learning as much as possible about it was imperative, held several international meetings during the last decade of the twentieth century and the first decade of the twenty-first and produced an impressive array of scrupulously detailed, highly specialized studies. ("Diogenes A: Diagnostic Observation of the Geology of Near-Earth Spectrally-Classified Asteroids" is one example among hundreds.)

The Association of Space Explorers International Panel on Asteroid Threat Mitigation decided in 2008 that, since the threat from NEOs is worldwide, the whole world should know about it and, if possible, get involved—and that obviously included the United Nations. The association itself was formed in 1983 by a large group of the space community's leading lights, including several who had been to space. The fact that it had a twenty-five-member panel that worked on trying to reduce the threat was a clear indication of how seriously its members took the problem. The panel also decided that it was important to alert the international community about the potential danger and the need to do something to protect the planet.

It therefore submitted a report, *Asteroid Threats: A Call for Global Response*, to the United Nations in September 2008. After providing the requisite background information, including Tunguska and 99943 Apophis, "which has a 1-in-45,000 chance of striking Earth in 2036," and explaining that such events are infrequent in a human lifetime but can be devastating when they do occur, the panel reiterated what had long been known about the situation and made the case for an international response to it. "Because NEO impacts represent a global, long-term threat to the collective welfare of humanity, an international program and set of preparatory measures for action should be established. Once in place, these measures should enable the global community to identify a specific impact threat and decide on effective prevention or disaster responses. A global, coordinated response by the United Nations to the NEO impact hazard should ensure that three logical, necessary functions are performed," the document stated.

The first function was information gathering, analysis, and warning, which, as everyone in the community had always agreed, meant knowing the enemy. The second dealt with mission planning and operations and called for a group having that title to assess the global capacity to deflect a hazardous NEO by specifying what technologies would be required and what the capabilities of participating space agencies had to be to deflect it. And lastly, the panel suggested that the United Nations oversee those functions with an intergovernmental Missions Authorization and Oversight Group, as it came to be known, that would develop policies and guidelines that represent the international will to respond to the threat and submit them to the UN Security Council. That presumed there was an international will and that the international community would act cohesively to address the potential danger with a specific defensive plan. It was a stretch that was tantamount to getting the community to agree on the definition of responsible government or what constitutes great art.

"The Association of Space Explorers and its international Panel on Asteroid Threat Mitigation are confident that with a program for concerted action in place, the international community can prevent most future impacts," the panel stated, carefully using the qualifier *most*. "The Association of Space Explorers and its international Panel are firmly convinced that if the international community fails to adopt an effective, internationally mandated program," the document warned, "society will likely suffer the effects of some future cosmic disaster— intensified by the knowledge that loss of life, economic devastation, and long-lasting societal institutions could have been prevented. Scientific knowledge and existing international institutions, if harnessed today, offer society the means to avoid such a catastrophe. We cannot afford to shirk that responsibility."[2]

The deliberative body deliberated for almost five years, and its Committee on the Peaceful Uses of Outer Space finally came up with Action Team 14, which issued a report—ironically, three days after Chelyabinsk—that called for international cooperation in the form of an International Asteroid Warning Network, an Impact Disaster Planning Advisory Group, and a Space Mission Planning Advisory Group. "If the proposed coordination mechanism was in place," Sergio Camacho, the chairman of Action Team 14, said at a news conference in Vienna, "then at minimum it would have allowed for more observation and better understanding and education of the population on what to expect rather than having a surprise effect with people not knowing what was happening."[3]

Given what had turned up at Chicxulub—both literally and figuratively—it became clear that there was no higher priority

for Earth than defense against objects that could cause the ultimate catastrophe. The Alvarez team's discovery, which was a
supremely important event in paleontology, was reported in an
article they wrote that appeared in the June 6, 1980, issue of
the respected and widely read journal *Science*. It started serious
discussions in universities and in intellectual circles, generally
about the extinction of the dinosaurs and many other creatures
and about the large object that did them all in.[4] If the dinosaurs
had had a space program, some in the aerospace community
chortled, they would still be here. (They had an air force of
pterodactyls, but it did not help.)

Congress eventually caught on and told NASA as much
in the space agency's 1991 authorization bill. "The House
Committee on Science and Technology believes that it is imperative that the detection rate of Earth-orbit-crossing asteroids
must be increased substantially, and that the means to destroy
or alter the orbits of asteroids when they do threaten collisions
should be defined and agreed upon internationally. The chances
of the Earth being struck by a large asteroid are extremely small,
but because the consequences of such a collision are extremely
large, the Committee believes it is only prudent to assess the
nature of the threat and prepare to deal with it."

NASA responded by issuing a Spaceguard Report in 1992
that laid out a plan for what would soon become known as
the Spaceguard Survey, a comprehensive cataloging of all
potentially dangerous NEOs. (Recall that, by then, the term
Spaceguard, as used by Arthur C. Clarke in *Rendezvous
with Rama* two decades earlier, had been cut in stone by the
planetary-defense community as a suitably descriptive term for
what was required.):

> Impacts by Earth-approaching asteroids and comets pose a signifi-
> cant hazard to life and property. Although the annual probability
> of the Earth being struck by a large asteroid or comet is extremely
> small, the consequences of such a collision are so catastrophic that

it is prudent to assess the nature of the threat and prepare to deal with it. The first step in any program for the prevention or mitigation of impact catastrophes must involve a comprehensive search for Earth-crossing asteroids and comets and a detailed analysis of their orbits. Current technology permits us to discover and track nearly all asteroids or short-period comets larger than 1 km diameter that are potential Earth-impactors. . . . What is required . . . is a systematic survey that effectively monitors a large volume of space around our planet and detects these objects as their orbits repeatedly carry them through this volume of space. . . . The international survey program described in this report can be thought of as a modest investment to insure our planet against the ultimate catastrophe.[5]

The document went on to describe in considerable detail what the space agency believed would be an adequate asteroid- and comet-surveillance program.

The Lowell Observatory in Flagstaff, Arizona, got into the NEO search in 1993 with funding from NASA, and it straightforwardly called the project the Lowell Observatory Near-Earth-Object Search. The facility is named for Percival Lowell, who founded it in 1894, and who was by all odds one of the most interesting characters in the history of American science. He was born in Cambridge, Massachusetts, in 1855, the scion of Brahmins who settled on the north shore of Cape Ann in 1639. He went to Harvard University, where he selected a challenging major: mathematics. It was a portent of the life that followed. Lowell was the personification of the Protestant ethic. He went on to become a businessman, a mathematician, an author, and an astronomer who titillated his colleagues by suggesting that there are canals on Mars that were built by an ancient civilization in a desperate attempt to survive. He also published a book, *Mars as the Abode of Life*, in 1908 that lent credence to the theory and undoubtedly helped H. G. Wells's *The War of the Worlds*, which was published ten years earlier, become an enduring work that spawned half a dozen feature films and an infamous radio drama on the Columbia Broadcasting System's Mercury Theatre on the Air that aired on October 30, 1938—the infamous Halloween broadcast. In said broadcast, actor Orson Welles played a reporter who excitedly narrated the Martian attack with ray guns as if it were an actual news bulletin. Thousands of listeners in the northeastern United States and Canada who tuned in late thought it was real and became so terrified that they poured out of their homes and onto the streets to see the hideous invaders for themselves. The creatures ultimately evolved into those little green men in flying saucers who occasionally visited Earth, a myth that was dispelled when Mariners 4, 6, 7, and 9 and the two Viking spacecraft got to the Red Planet in the 1960s and '70s and found no canals, let alone ray-gun-wielding Martians.

The Lowell Observatory has observed more than five million asteroids, including 289 that were near Earth, fifty-nine of them declared to be potentially hazardous. It also discovered forty-two comets. On a night of good seeing, as astronomers call clear, dark nights with excellent visibility, the system could locate as many as six thousand asteroids, a few comets, and perhaps three or four new NEOs. During its fifteen-year life, which began in 1993 and ended in 2008, the Lowell Observatory Near-Earth Object Search (or LONEOS, as the observatory was called) made about 450,000 exposures in 130,000 regions of the sky. That turned into about fifteen terabytes of data, in the arcane argot of computerdom. Some idea of the magnitude can be gained by knowing that if those bytes of data were words instead of parts of images, they would fill twenty million three-hundred-page books, which is about thirty times the printed material in the Library of Congress. But thanks to the marvels of computer miniaturization, all of it can be stored in two suitcase-size server computers.[6]

Whatever Tom Gehrels and Robert S. McMillan thought about *Tyrannosaurus rex*, pterodactyls, and other vanished reptiles, they were astronomers in the University of Arizona's internationally respected Department of Astronomy and they were therefore well aware that asteroids and, to a far lesser extent, comets infest the neighborhood. And they certainly knew about what the Alvarez team discovered in the Yucatan. So they decided that the NEO "population" had to be meticulously studied and cataloged—inventoried—and that new computers could handle that task. Starting in 1984, Gehrels led a team of astronomers that used the large Newtonian telescope at the Steward Observatory on Kitt Peak, Arizona, to monitor every asteroid and comet they could find. They christened their project "Spacewatch."

In 2001, Spacewatch began observing with a new telescope that had twice the aperture of the original one and that could therefore follow up on asteroids that become fainter after they are discovered. That gave them a more complete picture of asteroids' life cycles, which in turn helped to assess their danger. And to make matters even more interesting, since asteroids and comets are Solar System debris, the increased knowledge about them contributed to an understanding of how the Solar System was formed and how it works. Serendipity! McMillan was Spacewatch's principal investigator and went on to head the organization, which discovered some asteroids, perhaps most notably 20000 Varuna, an egg-shaped rock in the Kuiper Belt that some astronomers think is large enough to qualify as a dwarf planet. Like Kali, Varuna was named after a Hindu god, but this one a benevolent type who presided over the waters in heaven and was the guardian of immortality. Spacewatch's main contribution was not finding asteroids that came in close but in demonstrating the technology that was required to do it.[7]

The Catalina Sky Survey watches "up over" as well as "down under," which is to say it uses large telescopes on the peak of Mount Lemmon and on Mount Bigelow, both in the Tucson area, and one at Siding Spring Observatory in Australia.

Many of those who search the sky for NEOs and potentially hazardous asteroids (PHAs) are University of Arizona Wildcat Astronomers, whose offices and classrooms are nearby. The Catalina group has turned up more NEOs and PHAs than any other observatory, including Lincoln Near-Earth Asteroid Research (LINEAR), with 310 NEOs discovered in 2005, 396 in 2006, 466 in 2007, and 564 in 2008.

The Near-Earth Asteroid Tracking Program, which went by the public-relations-friendly acronym NEAT, did the same thing at the Jet Propulsion Laboratory (JPL) between 1995 and 2007, when it discovered thousands of asteroids using the US Air Force's advanced Ground-based Electro-Optical Deep Space Surveillance system on the Hawaiian island of Maui that was designed to scrutinize Soviet and other spacecraft during the Cold War. (The size of the antenna of a satellite that is positioned to intercept radio and other communication "traffic" on Earth gives a good indication of its capability, and the altitude of imaging reconnaissance satellites—spy satellites, as they are commonly called—provides an excellent clue as to what they are taking pictures of, since their altitude is lowered before their imaging systems are turned on and then raised after the "target" has been imaged. There are things to be learned just by knowing where a spacecraft is and what it looks like.)[8] It also used a large telescope at the Palomar Observatory in California. NEAT not only turned up thousands of new asteroids, but it also discovered or codiscovered several important comets. And whatever else it accomplished, NEAT had the distinction of being a contender for honors as a Great Acronym of Distinction (GAD), along with the American Indian Movement (AIM); Acquired Immune Deficiency Syndrome (AIDS); Zone Improvement Plan (ZIP-code); Near-Earth Asteroid Rendezvous (NEAR); and Light Amplification by Simulated Emission of Radiation (LASER).

Then came Sentinel, a decade-old concept that literally added a dimension to planetary defense by in effect stepping way back and spotting NEOs in the infrared, so they stand out against the cold sky around them. The spacecraft's development was the brainchild of the B612 Foundation, and specifically of Russell L. "Rusty" Schweickart, Clark Chapman, Ed Lu, and Piet Hut. On Thursday morning, June 28, 2012, the foundation held a news conference at the California Academy of Sciences in San Francisco, at which it announced an important strategy for locating potentially hazardous asteroids. "The orbits of the inner Solar System where Earth lies are populated with a half million asteroids larger than the one that struck Tunguska, and yet we've identified and mapped only about one percent of these asteroids to date," Lu, who had become chairman of the foundation, told journalists. The conference was called to announce the foundation's plan to build, launch, and operate Sentinel, the first privately funded spacecraft to go on a deep-space mission that, by definition, is far from Earth. The B612 Foundation estimated that the cost of developing and launching Sentinel would come to roughly $450 million.[9]

The concept was not new, but its execution was. Instead of continuing to peer out from Earth in a necessarily narrow and limited field of vision to see what might be menacing it, or scanning the sky from a spacecraft orbiting Earth, B612 proposed to step very far back and see the big picture as it looks from a Venus-like orbit around the Sun. "We believe our goal of opening up the solar system and protecting humanity is one that will resonate worldwide," Lu said. "We've garnered the support and advice of a number of individuals experienced with successive philanthropic capital campaigns of similar size or larger, and will continue to build our network. We've been given a gift, and the gift is that we have the ability now to go out there and actually do something which positively affects the future of humanity on Earth."[10] A telescope in a Venus-like orbit would in fact not be efficient watching Earth's immediate neighborhood and would not see the planet at all half the time.

Sentinel's assignment is to locate 90 percent of asteroids 140 meters or larger in diameter that are in the inner Solar System and report what it finds. The foundation estimated that there are roughly half a million asteroids in the vicinity of this planet that are as large or larger than the one that exploded over Tunguska. The telescope, which will be made by Ball Aerospace and Technologies Corp. of Boulder, Colorado, will be 7.6 meters—twenty-five feet—long and will see in the infrared (again, so it can pick out relatively warm rocks against the very cold background of space). Sentinel is scheduled to be launched on a two-stage Falcon 9 rocket that was manufactured by the Space Exploration Technologies Corp., or SpaceX, a company in Hawthorne, California. SpaceX was started in 2002 by Elon Musk, a very imaginative entrepreneur and hyperactive inventor who has revived an old science fiction transportation mode for California: a so-called Hyperloop express train that would carry passengers across California at more than 1,200 kilometers an hour, making the ride from San Francisco to Los Angeles in a little more than half an hour. That's faster than by jetliner. And it would certainly reduce freeway congestion.[11] Falcon 9 is a model for the kind of innovative and aggressive enterprise that is commercializing space after the shuttle's retirement. SpaceX also produced a partially reusable spacecraft called *Dragon*. The firm was awarded a $1.6 billion NASA contract in 2008 for twelve resupply flights to the International Space Station in which *Dragons* would be launched by Falcon 9s. The first one was successfully flown in October 2012. But the Sentinel mission stood to be of far greater importance where civilization's survival is concerned. Thus did the formidable founder of SpaceX become involved with B612 and planetary defense as well as all of his other lofty projects.

Pan-STARRS—for the Panoramic Survey Telescope and Rapid Response System—is an array of telescopes and astronomical cameras that survey the sky continuously and send what they find to a central computing facility at the University of Hawaii's Institute for Astronomy (which, like the University of Arizona, is a veritable incubator of space watchers). It has financial support from an international consortium that includes Johns Hopkins and Harvard, with the US Air Force funding four 1.8-meter telescopes, the first of which began operating full-time at the summit of Haleakala on Maui in May 2010. Pan-STARRS's essential purpose is to spot asteroids and comets by comparing successive pictures so that anything that has moved stands out. The system was started as a collaboration between the Institute for Astronomy, LINEAR, the Maui High Performance Computing Center, and Science Applications International Corporation (better known as SAIC), the last of which is a defense contractor and think tank in Tysons Corner, Virginia. That location, uncoincidentally, puts it near CIA headquarters at McLean and on the same side of the Potomac as the Pentagon (and its procurement office).

ATLAS, which is as venerable a name as there is in the English lexicon, stands for the Asteroid Terrestrial-impact Last Alert System Project—another neat . . . acronym—that is being developed at the University of Hawaii for use starting in 2015. All the scientists in the project have worked at Pan-STARRS, and ATLAS is intended to compliment it by searching "shallow but wide" sections of space rather than smaller "deep but narrow" portions of it with great magnification as Pan-STARRS does. The problem with Pan-STARRS, at least in the opinion of the ATLAS group, is that asteroids and comets can sometimes streak by while the telescopes are looking in a different direction. ATLAS is therefore designed to use as many as eight small telescopes with ten- to twenty-inch lenses at two or more locations to scan the entire visible dark sky twice a night. Ideally, its designers have said, it will operate in California and Arizona as well as in Hawaii. There are two advantages to spreading out the telescopes. Coverage at three locations will triangulate asteroids and comets—see them from three directions—which will not only make fixing their positions more accurate but will also enable the system to see smaller NEOs than a one- or two-dimensional operation could. It will also allow the sky to be watched constantly so that clouds or storms will be far less likely to blind the system. Spaceguard actually covered that territory in its 1992 report, but duplication in planetary defense is an asset, not a sin.

John Tonry, the astronomer who heads ATLAS, compared its sensitivity to seeing a lighted match in New York from San Francisco. He of course knows that the object of planetary defense is to spot the object so far ahead of a collision that it can be diverted, but his group concentrates on the last-ditch scenario; on the asteroid or comet that is not a world-ender and that connects. The team predicts that there will be a week's warning time for fifty-meter "city-killers" and three weeks' warning time for 150-meter "county-killers." The reaction plan that he describes brings chillingly to mind the civil-defense procedures that were put in place in case of a nuclear attack during the Cold War. "That's enough time to evacuate the area of people, take measures to protect buildings, and other infrastructure, and be alert to a tsunami danger generated by ocean impacts," he added.[12]

The Cold War analogy is not far-fetched. The project measures the impact threat the way the devastation caused by nuclear weapons is measured, so the "town-killers" are thirty kilotons, the "city-killers" are five megatons, and the "county-killers" are one hundred megatons. "Yes, these are the same kilotons and megatons of TNT used to describe nuclear explosions because a serious asteroid impact is very similar to a nuclear explosion, including a mushroom cloud, but without radioactivity or fallout," the ATLAS group has explained. Ultimately, they want planetary defense to operate "robotically," meaning automatically, and grow into a network that extends to the

Southern Hemisphere—Australia, Chile, or South Africa are mentioned—so there will be constant 'round-the-planet coverage. There would always be the biggest picture possible, in other words, and it would be a matter of staring at that picture all the time electronically to see what moves and, more to the point, what moves as a menacing potential impactor.

The Wide-field Infrared Survey Explorer, or WISE, as it is called by the planetary-defense fraternity, was another astronomical space telescope that was also designed to survey the sky in the infrared. It was launched into a polar orbit in mid-December 2009, and, by the following October, it had discovered more than 33,500 new asteroids and comets, including 136 near-Earth asteroids (NEAs), PHAs, and some comets that, typically, were not menacing. That is a tiny fraction of the roughly ten thousand that have been discovered by telescopes on terra firma. Nineteen of the PHAs were believed likely to hit Earth at some point and cause significant damage, in which case they will realize their potential unless they are stopped. That same month, the hydrogen coolant that kept the temperature of the telescope down was depleted, but the spacecraft had sent back so much data that it was rechristened NEOWISE, an evident play on NEO's two meanings—"near-Earth object" and "new"—and its mission was extended four months. The spacecraft then imaged roughly six hundred NEAs, leading its team to conclude that there are roughly 4,700 PHAs with diameters larger than one hundred meters that were relatively near Earth. The new observations also suggested that about twice as many PHAs are likely to be in lower inclination orbits, which are more aligned with Earth's orbit, than was previously thought.

"NASA's NEOWISE project, which wasn't originally planned to be part of WISE, has turned out to be a huge bonus," Amy Mainzer, the mission's principal investigator at JPL, said. "Everything we can learn about these objects helps us understand their origins and fate. Our team was surprised to find the over-abundance of low-inclination PHAs. Because they will tend to make more close approaches to Earth, these targets can provide the best opportunities for the next generation of human and robotic exploration," she said in an apparent reference to landing a human or two on one of the larger boulders.[13]

Lindley Johnson, who studied NEOs intensively in the air force and then went to NASA, where he became the program executive for the Near-Earth Object Observation Program and a recognized expert in the field, agreed about the importance of extending WISE.

> The NEOWISE analysis shows us we've made a good start at finding those objects that truly represent an impact hazard to Earth. But we've many more to find and it will take a concerted effort during the next couple of decades to find all of them that could do serious damage or be a mission destination in the future.
>
> We think we've got a pretty good handle on anything that's large enough to cause global consequences. Our estimate is that we've found somewhat over 95 percent of those that are large. There's always going to be a few lurking out there in deep space that we will pick up as we continue the survey of the smaller objects. Our current objective is to find all the hazardous asteroids down to about 100 meters. That is a considerably more challenging mission than the one-kilometer size. We are still using the same telescopes that we used for the larger size mission because to date we have not gotten the funding to really be doing an adequate job for the smaller size missions. We have to have either a lot bigger telescopes on the ground for decades or you've got to get a space-based capability.[14]

Johnson's complaint was a familiar one in Washington. NASA's NEO program is so important that it warrants being a permanent, integrated, adequately funded operation, but its funding does not reflect that. "You've got to find them first. We're trying to design a program that will find them years in advance so that we do have time to do something about it. It's perfectly within our technical capability to be able to find all of those of a significant size that could have impact potential.

We've just got to do that." But the effort lacks the funding to be as focused as it should be, he continued. "NASA's budget right now is well over-subscribed. We have a lot more on our plate to do than we have been given the budget to accomplish. This is something that you've got to anticipate well in advance and be looking at now to find an impactor that might occur in the second half of the century. It's the kind of lead time you prefer to have in dealing with something like this, particularly if it's a large object. We prefer to be able to intercept it years in advance and just give it a slight nudge and not have to worry about it."

And, like in *The Little Prince*, Lindley Johnson thinks that asteroids have a good side as well as a bad one. "As we explore the Solar System and move out into it, how to live off the land, so to speak, will become important, and so mining asteroids, finding resources on them, certainly including water, will be important to our future in the Solar System." He also stressed cooperation with the European Space Agency (ESA) and other foreign space organizations "to determine and then verify results before we make predictions."[15]

International cooperation to protect the international community is a good deal less than it should be, but there are indications that it could strengthen. Although the United States leads the world by far in finding and cataloging asteroids and comets, threatening and otherwise, some other political entities have come to understand that since the danger is worldwide, so must be the response. Italy and Spain therefore started NEODyS, for Near-Earth Object Dynamic Site, in 1999. It is a program that calculates the orbits of asteroids and provides updated information on the Internet about where they are and where they are expected to be through 2100. The site is run by the University of Pisa. It's main feature is a risk page on which all NEOs that could theoretically hit Earth through the end of the next century are shown on the list, which grades them in five categories of probability, ranging from those that are lost or are too small to worry about to those in a "special" category that are considered to be candidates for a possible impact (99942 Apophis is prominently included).

ESA started the Near-Earth Object Coordination Centre at the European Space Research Institute in Frascati, outside of Rome, on May 22, 2013. The centre operates under ESA's Space Situational Awareness Program, which began in 2009 and is developing an independent capability to watch for natural phenomena that could disrupt satellite operations in orbit or vital operations on Earth, such as power grids and communication. It consists of three main segments: monitoring conditions at the Sun, in the solar winds, and in Earth's space environment that can affect the planet's space- and ground-based infrastructure or endanger human life or health; watching for active and inactive satellites, spent launch stages that are in orbit, and all the space debris left over from the world's combined launches; and, last but by no means least, detecting NEOs that are potential impactors. So far, in common with other governmental agencies of all stripes the world over, it has not detected anywhere near the number of NEOs it has declared it wants to detect. It should be noted that the United Kingdom was only the second nation after the United States to endorse NEO surveys and planetary defense in an excellent 2000 parliamentary report.

The NEO Coordination Centre did have a clear triumph on June 26, 2013, though, when astronomers in several observatories spotted and sized up asteroid 2002 GT as it passed the planet at a distance of almost fifty times the distance to the Moon and coordinated what they saw. At that distance, 2002 GT posed no threat, but at a few hundred meters across, it made an excellent practice subject for any others in its size range that get closer. "The fly-by presented an ideal opportunity to exercise the unique coordinating function of ESA's new Near-Earth Object Coordination Centre," Ettore Perozzi, the project leader for NEO services at Deimos Space, declared in a tone of triumph that was somewhat exaggerated given the centre's record. "By alerting and then collating observations from diverse European teams, the Centre was able to provide a comprehensive set of results back to the scientific and space exploration communities, a cycle that wasn't happening before. This is really a first for Europe," he proclaimed. The Spanish company operates a complete remote-sensing system with a satellite named Deimos-1. In keeping with the space community's penchant for dark humor, Deimos is the son of Ares and Aphrodite and the twin brother of Phobos in Greek myth. The name literally means "dread" and was the personification of the sheer terror that is brought on by war. The two moons that orbit Mars, which is itself named for the Roman god of war, are named Deimos and Phobos, and in an irony that Shakespeare would have savored, are thought to be captured asteroids.[16]

The far encounter with 2002 GT delighted astronomers and others in the space-science world because the asteroid is the intended target in a mission called Epoxi, in which a probe named Deep Impact, which was launched in 2005 (and which flew by and looked over comet Hartley 2 at close range and studied the composition of comet 9P/Tempel by sending an impactor into it) is supposed to intercept 2002 GT and scrutinize it, NASA's fragile budget permitting.

"Traditionally, Europe's asteroid community reliably delivered world-class observations and has been credited with many significant discoveries and findings," Gerhard Drolshagen, the comanager of the NEO segment in the Space Situational Awareness Program said. "What was lacking, however, was a central point to coordinate and synthesize data that could function across national and organizational boundaries. Our centre has proven it can act as a driving force and a focal point for the European and international community involved in asteroid science, impact monitoring and mitigation."[17]

Japan has its share of asteroid watchers (and worriers), too.

The Japan Spaceguard Association was established in 1996 to learn more about the threat, and what it learned convinced its members that they should be on the lookout for potential troublemakers, along with other nations that call themselves advanced. After long and arduous discussions with the government, the association finally won a concession to build an observatory at Bisei, a town in the low mountains of the Okayama Prefecture 150 kilometers west of Osaka. It was built in 1999 with funding from the Japan Aerospace Exploration Agency, is called the Bisei Spaceguard Center, and is the heart of the Japan Spaceguard Association. It claimed to have discovered more than one thousand asteroids, two of them NEOs, and a comet as of early February 2012.[18]

But the Japanese have been more than passive in addressing the situation. They have sent a probe to land on an asteroid, as the NEAR mission did, but with a very significant difference: it brought a sample home. The spacecraft was named Hayabusa (peregrine falcon) and was sent by the Japan Aerospace Exploration Agency to land on an asteroid named Itokawa in May 2003. It spent two and a half months flying with the rock and then landed on it in November 2005. Hayabusa not only sized up Itokawa pretty thoroughly, collecting data on its size, shape, spin, topography, color, mineral composition, density, and gravity, but it also picked up tiny grains of, well, asteroid and brought them home in June 2010.[19]

A spacecraft belonging to the China National Space Administration, on the other hand, had a dramatic encounter with an asteroid while on a lunar mission. The probe, Chang'e 2, was launched in October 2010 to collect data on the Moon as part of the first phase of China's low-key but persistent lunar-exploration program that was planning an eventual soft landing on the Moon by a rover called Chang'e 3. Its predecessor, Chang'e 2, went into lunar orbit in late August 2011 and transmitted data home until mid-April 2012, when its controllers ordered it to leave the Moon and head for 4179 Toutatis, a 4.5-kilometer-long asteroid with two distinct "lobes" that came within 585,000 kilometers of Earth—that is four times the distance to the Moon—on September 29, 2004.[20] The asteroid has made several relatively close approaches to this planet, is due back in 2016 and 2069, and is listed as a potentially hazardous object even though the probability of its orbit intersecting Earth's is essentially zero for the next six centuries.[21]

The grandiose plan to land taikonauts on the Moon, which reflected a determination to be recognized as the third superpower at the end of the twentieth century, is hardly mentioned during a time of economic problems and regional competition with Japan over tiny, disputed islands in the East China Sea. Beijing's interest in space, reflecting that pragmatism, is primarily military and commercial. The military dimension became clear when the People's Liberation Army blinded US reconnaissance satellites with ground-based lasers in 2006 and obliterated an aging weather satellite in an antisatellite test on January 11, 2007. That got the attention of the satellite-dependent US military and intelligence communities. In March 2013, Beijing announced that it was combining four maritime units into one superagency that a senior Chinese navy official described as an "iron fist."[22] And economic imperatives being what they are, China has concentrated its civilian space program overwhelmingly on carrying other countries' payloads—communication and weather satellites, for example—to orbit for profit. It can be deduced that the men who rule China from inside the secluded Zhongnanhai have decided to leave planetary defense to other nations and concentrate their country's resources on national, not international, defense in the most expansive meaning of the term; a healthy economy being crucial to national defense.

Having experienced Tunguska, Russians needed no reminder of the danger posed by NEOs, but Chelyabinsk brought it home yet again. So, barely four months after the explosion, on June

25, 2013, Vladimir Puchkov, the nation's emergencies minister, announced at a news conference in Moscow that his country and the United States would work together to improve protection against meteorites and other threats from space. "We have decided that the U.S. Federal Emergency Management Agency and Russia's Emergencies Ministry will work together to develop systems to protect people and territory from cosmic impacts," he said. "I believe we can make a technological breakthrough in this area if the Emergencies Ministry and FEMA supervise this project, attracting the finest minds and research groups, including in Canada, Europe, China and Southeast Asia."[23]

Roskosmos, the Russian space agency, announced almost simultaneously that it intended to develop a system of automated telescopes on Earth and in space that will warn of impending danger. And the system would be passive-aggressive, which is to say that an intercontinental ballistic missile would carry a small attack spacecraft to space and then launch it at any object that is confirmed to be a threat. Deputy Chief Designer Sergei Molchanov of the Makeyev State Rocket Center announced at the annual Korolev Readings forum in Moscow in January 2013 that the missile would launch a "destroyer" spacecraft known as a *kapkan* at the approaching impactor and either hit it head-on or push it off course. Hitting the NEO head-on would, in theory, be like launching an antiballistic missile (ABM) at an incoming warhead, which the United States has tested over the Pacific Ocean for years with a success rate of 50 percent. That is considered wholly inadequate given the consequences of a successful attack. But there would be a crucial difference between ballistic-missile defense and NEO defense. The ABM system has perhaps an hour to stop a warhead or a cluster of them, whereas the destroyer or its international counterpart would have months or hopefully years to stop the impactor.[24]

Two months later, on March 12, the head of Roskosmos, Vladimir Popovkin, announced that Russia intended to build a shield to protect Earth from meteors and other things that come out of the sky as part of an international project that he said was called Citadel and that would cost some $500 million. "Roscosmos has formed a working group with experts from the defense ministry and the Russian Academy of Sciences to create a unified system of early warning and countering space threats," he said.[25]

In announcing that Citadel was to be an international project, Popovkin did not seem to take into account the international defense plan that calls for a graduated response. Furthermore, Roskosmos had a distinctly poor record in following through with its NEO defense plans. Popovkin told legislators that a rough plan to deal with the asteroid threat would be in place by the end of 2013; Oleg Shubin, the deputy director of the nuclear agency's munitions-experiments division, said that intercepting an asteroid that is a kilometer or larger will definitely require a nuke. But the asteroid problem soon became moot, at least where Popovkin was concerned. Within five months of the announcement of the destroyer program, he was officially reprimanded by none other than Dmitry Medvedev, the prime minister, for incompetence because of a series of launch failures that embarrassed the Kremlin. Medvedev was a high-ranking official in the nation that opened the Space Age with *Sputnik* and followed it with Gagarin's and Tereshkova's flights and all the rest, so the fact that the launch failures angered him was entirely understandable, particularly with the ESA and Chinese programs in the ascendance (in both senses of the term).[26] The incompetence extended well beyond the launchpads at the Baikonur Cosmodrome, near Tyuratam, on the desolate steppe in Kazakhstan, east of the Aral Sea. Roskosmos was being scorned internationally for coming up with elaborate planetary-defense plans that never seemed to materialize, and Popovkin took the hit for that, too.

The American Institute of Aeronautics and Astronautics

(AIAA), which was started in 1963 with the merger of the venerable American Rocket Society and the American Interplanetary Society, took its place in planetary defense in 1990, right after Apollo Asteroid 1989FC passed within four hundred thousand miles of Earth—a veritable near miss in cosmic terms—and took the planet's inhabitants by surprise.[27] That rock was larger than an aircraft carrier and crossed Earth's orbit at a point where the planet had been only six hours earlier. Had there been a collision, it was later calculated, it would have released the equivalent of as many as 2,500 one-megaton hydrogen bombs. And an impact in a large metropolitan area with high population density would have killed millions of people instantly without their even knowing what had happened. So, in 1990, the AIAA published a position paper, "Dealing with the Threat of an Asteroid Striking Earth," which called for a systematic and open program to detect and define Earth-crossers so precisely that impacts can be predicted with some confidence. It went on to recommend that a study be done to specify the systems that would be able to detect, destroy, or significantly alter the orbits of those that are believed to be on a collision course with this planet.

The paper also suggested that the technology that was developed for US and Soviet space programs (it specifically mentioned the Strategic Defense Initiative) be studied for use against asteroids, although they would be used at far greater distances than they would have been to intercept ballistic missiles. The distance factor would be particularly important if the asteroid-busters are nuclear, which will most likely be the case (more about that further on). Even though it was published when the "giggle factor" was occurring in films and in other media, as was described in the fourth chapter, the paper was referred to more than two thousand times by 1993, making it the most cited AIAA position paper to that time.[28] It was followed by one in 1995 that made the same recommendations and that was very likely responsible, at least in part, for the cre-ation of the Near-Earth Object Program Office.

The association took matters a giant step further when it held the first Planetary Defense Conference: Protecting Earth from Asteroids, in Anaheim, California, on February 23–26, 2004. The meeting was historically important because it was an incubator in which most of planetary defense's nationally recognized stalwarts convened for the first time to present their work and get acquainted in a synergistic atmosphere. The par-ticipants included Andrea Carusi of the European Space Agency; Clark Chapman; Lindley N. Johnson of the Planetary Science Institute; John Logsdon of George Washington University; Ed Lu; Rusty Schweickart, then of the B612 Foundation; David Morrison, an astrophysicist and the director of space at the NASA Ames Research Center; Steven J. Ostro, a radar astron-omer at JPL who was part of the team that discovered 222 NEAs, 130 of them that were potentially hazardous; Donald K. Yeomans; and Simon P. Worden, a retired air force brigadier general and a research professor of astronomy at the University of Arizona, who was a congressional fellow.

The meeting produced a white paper that summarized its conclusions, including recommendations to review interna-tional professional and amateur efforts to detect potentially threatening asteroids and comets and improve coordination—communication—among them, survey and catalog NEOs in the one-hundred-meter class, encourage the development of cre-ative ideas for finding and cataloging potentially threatening long-period comets, develop and fund ground-based observa-tion techniques and missions to the asteroids themselves, and establish a formal procedure for getting information out when the probability of an impact exceeds specified thresholds.

"This conference is the first of a series of the threat posed by Near Earth Objects, possible techniques and missions for deflecting an oncoming object, and political, policy and disaster-preparedness issues associated with NEO deflection," the paper said in summary. "The conference produced several recommended actions, the foremost being that we need to: 1) begin trust-building efforts so that claims that the NEO hazard is important will be considered credible by the public, even though we recognize that the probability of a disastrous impact is small; 2) increase our efforts to detect threatening objects and to determine the detailed physical and compositional properties of NEOs; and 3) move forward on means to deflect a threatening object. A key recommendation, consistent with previous AIAA Position Papers, is that a chain of responsibility be clearly and publicly defined for detecting and warning the public of threats, and mitigating those threats. These threats are real, and efforts to coordinate information and activities related to detecting and mitigating them should begin now. . . . Future impacts by comets and asteroids are a certainty," the paper concluded. "Such impacts could have severe consequences—even ending civilization and humanity's existence. Life on Earth has evolved to the point where we can mount a defense against these threats. It is time to take deliberate steps to assure a successful defensive effort, should the need arise."[29] It is indeed time.

The gravity of the situation . . . impelled the AIAA to hold a second Planetary Defense Conference from March 5–8, 2007, that one at George Washington University in the nation's capital, very likely with the hope that it would draw reporters from the *Washington Post* and the *New York Times* and, in the process, get some attention on Capitol Hill. That meeting also focused on detecting, characterizing, and mitigating NEOs, and there were also presentations and discussions on the political, legal, and societal challenges that would affect mounting a defense. As was the custom, the first day was, for the most part, spent describing the situation, including defining PHOs—potentially hazardous objects—that are fifty meters or larger and describing them. The second day was devoted to techniques that could be used to deflect them or hit them head-on, breaking them into small fragments. The third day was about the likely consequences of impacts, such as tsunamis and overpressure from airbursts. A panel also considered legal issues that would arise in the testing and implementing of deflection, along with that, it considered ways to maintain funding for planetary protection. The last day was spent discussing the international decision-making process, which would necessarily depend on developing a working relationship within the international community and a consensus on what needs to be done.

"While significant scientific and technological advances have been made since the 2004 conference and are ongoing," the concluding statement said, "it is clear that providing effective planetary defense from Near Earth Objects and planning for mitigation of an impact are in their infancy." And the meeting's primary conclusions were articulated. The participants agreed with others who look for and are appropriately wary of NEOs that most civilization-ending, kilometer-size asteroids and comets have been located but that smaller ones in the 140- to 300-meter range could strike without warning and cause "serious loss of life and property over a broad area." Earth-based radar antennas, such as the deep dish at Arecibo in Puerto Rico, are critical for providing the information that is needed for deflection which, the white paper noted, is still only in the conceptual phase. It also pointed out that there are serious technical and political issues, among others, in deciding whether and how to respond to a threatened impact and that a threat has never been seriously considered by any agency that would bear responsibility for dealing with it. Finally, the conference members went on record as noting, yet again, that the NEO threat is international and that it therefore demands international cooperation.

"Given the global nature of the consequences, it is unlikely that one country will decide on its own whether to take action,"

the document stated. "There must be international involvement in decision-making and in whatever actions are decided.
Discussions on how these decisions will be made should begin
while there is no specific threat. Principles and protocols for
the process of communication and dissemination of information about potential impacts, and the implementation of necessary mitigation measures should be negotiated and agreed
to at an international level. These protocols should identify
roles and responsibilities of key players and include a means to
notify governments and the public of all hazards of a regional
or global nature."

The mitigation measures were discussed on the second day,
when those who made presentations again emphasized that the
first requirement is knowing what the attacker is made of and
then deciding whether slow-push or quick-impulse deflection
would be most effective. It was put on the record that missions
to do them have been worked out and could be tested.[30]

At least as early as 1995, the AIAA was calling for an accelerated search for asteroids and short-period comets and the development of plans to stop them. In a position paper called "Responding
to the Potential Threat of a Near-Earth-Object Impact" issued
that year, it quoted Rep. George E. Brown Jr. of California, the
chairman of the Committee on Science, Space and Technology and
a staunch space advocate, as telling the following to a congressional hearing on the NEO threat in March 1993:

> If some day in the future we discover well in advance that an
> asteroid that is big enough to cause a mass extinction is going to
> hit the Earth, and then we alter the course of that asteroid so that
> it does not hit us, it will be one of the most important accomplish-
> ments in all of human history.

The position paper said that the AIAA strongly believed that
Brown's statement was correct and concluded with a warning.
"If some day an asteroid does strike the Earth, killing not only

the human race but millions of other species as well, and we could have prevented it but did not because of indecision, unbalanced priorities, imprecise risk definition and incomplete planning, then it will be the greatest abdication in all of human history not to use our gift of rational intellect and conscience to shepherd our own survival, and that of all life on Earth."

That being the case, the paper recommended the immediate approval of a program to accelerate the discovery, identification, and characterization of NEOs. Congress approved it three years later, and it became the Spaceguard Survey. The paper also recommended that a study be undertaken immediately to examine various concepts of responding to collision threats in the next century. "In the future," the AIAA concluded, "the U.S. should consider establishing an office for coordinating the U.S. response to this risk and should invite other nations to participate. The objective of this office is to provide the focal point for overall program management, planning and systems engineering, as well as coordinate delegated responsibilities regarding NEO detection, intercept, rendezvous, command and control systems and activities without international partners."[31] That office should be called the Department of Planetary Defense.

Meanwhile, where NASA is concerned, the proverbial handwriting is on the wall. Having had its manned space program suffer a setback by the retirement of the shuttles and the decision to abandon a return to the Moon, at least in the foreseeable future, and having successfully completed the Spaceguard Survey, the space agency has also come to believe that it has a serious role to play in planetary defense. It is working to find and inventory 90 percent of all NEOs 140 meters or larger by 2020 (within its existing and projected budgets). The space agency responded to *Defending Planet Earth: Near-Earth Object Surveys and Mitigation Hazards*, the National Research Council's 2010 report that described the situation and suggested mitigation possibilities, by reporting that it was already taking "a significant role" in plans for dealing with the NEO hazard in the UN's Committee on Peaceful Uses of Outer Space and in the international planetary-defense conferences. That was appropriate, since NASA commissioned and paid for the NRC study. NASA accepts the generally agreed-upon stages of collision avoidance: slow push, kinetic impact and, if they fail, the use of a nuclear weapon.

Ironically, nuclear weapons, which have been associated with what former secretary of defense Robert McNamara called Mutual Assured Destruction, could in fact assure survival. Bong Wie, the director of the Asteroid Deflection Research Center at Iowa State University, studied responding to a threatening asteroid on relatively short notice, which is to say a year or so, and concluded that a nuclear explosion is probably the only way to stop a large one in so relatively short a time. He came up with a hypervelocity asteroid impact vehicle that would hit the asteroid so hard that a crater would be formed and then a follow-up mission would plant a nuclear "device" in the crater that would set off the most efficient subsurface explosion possible and force the thing off course.[32]

The International Academy of Astronautics (IAA), which was established in Stockholm in 1960 by the legendary rocket pioneer Theodore von Karman, along with others who were committed to expanding the space frontier (as they put it), became so concerned about the NEO problem that it, too, joined the fray by extending a series of information-sharing planetary-defense conferences to Los Angeles in 2004; Granada, Spain, in April 2009; Bucharest, Romania, in May 2011; and a fourth in Flagstaff, Arizona, in April 2013. The meeting in Flagstaff opened with a session on Chelyabinsk, which was described as a "wake-up call." The IAA issued a white paper after Flagstaff that, like the AIAA and other groups, called for increased international cooperation and communication and the discovery, characterization, and movement of the NEO, as well as mitigation and constant preparedness. It also called for paying closer attention to the smaller asteroids, upgrading the world's space radars, considering sending low-cost probes to size up asteroids, and performing a kinetic impactor flight demonstration— that is, clobbering an asteroid to see what happens.[33]

While Congress has generally been supportive of NASA missions, in July 2013 it came out against one that would send an unmanned spacecraft out to capture a small asteroid in 2021 or afterward for close examination because it was considered frivolous, at least by the science committee in the Republican-controlled House. It was the centerpiece of the Obama administration's space-exploration agenda, but the Republicans laid out a plan that called for the space agency to send astronauts back to the Moon, set up a base there, and then go to Mars (on the cheap, with less money than NASA requested), not "lasso" a seven-to-ten-meter rock. "A costly and complex distraction" is how Rep. Steven Palazzo of Mississippi described the mission. Some of his colleagues complained that it seemed far-fetched and poorly articulated and, getting to the point, that it would not advance America's bragging rights in space the way a return to the Moon would. They were in a distinct minority, though. Congress was generally supportive of programs that contribute to planetary defense, and it remains so.

Obama had asked NASA to find a way to send astronauts to an asteroid by 2025 and then to Mars. NASA readily accepted the assignment and noted that it would "protect our planet" from dangerous asteroids in addition to making strides in human spaceflight. In June, the month before the Republicans laid out their objections, the space agency showed how concerned it was about the NEO situation by announcing its Asteroid Grand Challenge in which individuals and organizations were invited to find asteroids that threaten Earth and propose ways to end the threat. The space agency later reported that it received more than four hundred responses, including offers to help with a mission to capture an errant rock. And the profit motive in capturing one or more asteroids remained. At least two companies announced their intention to mine them for precious metals. But ultimately, even with the explosion over Chelyabinsk fresh in the lawmakers' minds, the asteroid and comet threat remained a formidable problem yet not as important as the manned program. A return of astronauts to the Moon and then striking out for Mars, even with a public that has largely lost interest in having people in space, still captivates the imagination of many who identify with human adventure, whereas the mundane business of preventing wandering rocks and chunks of dirty ice from hitting Earth seems almost irrelevant.[34]

Those who were aware of the danger, however, steadfastly believed that a real, workable, strategic plan, not theatrics and incessant hand-wringing, was required. NASA's Advisory Council Ad-Hoc Task Force on Planetary Defense was therefore created in 2010 to come up with a clear definition of what the NEO problem is and to devise a workable defensive scenario once and for all. It was cochaired by Rusty Schweickart, then of the B612 Foundation, and Thomas D. Jones, another former astronaut who was pedigreed as a Distinguished Eagle Scout and had a doctorate in planetary science from the University of Arizona, as well as four space shuttle missions to his credit (in addition to *Sky Walking: An Astronaut's Memoir*, a readable account of his career that made a good case for supplementing Earth's natural resources by mining asteroids, not shunning them). And there were five other luminaries on the planetary-defense task force: Clark Chapman; Donald K. Yeomans; Richard P. Binzel, a professor of planetary science at MIT; Lindley Johnson, then of NASA's Near-Earth Object Observation Program in Washington; and Brian Wilcox, a principal investigator of robotic-vehicle development for planetary exploration at JPL, planetary exploration's home roost. The group met in Cambridge, Massachusetts, in April 2010; it met again in Boulder, Colorado, that July; and members had two days of teleconferencing and Internet communication in August 2010 to discuss the planetary-defense situation and what NASA and the rest of the space community should do about it. The task force issued its final report on October 6–7, which recommended the following for the space agency:[35]

- Establish an organizational element to focus on the issues, activities, and budget necessary for effective planetary-defense planning, including a planetary-defense-coordination office, and challenge the international community to join analytical, operational, and decision-making activities. It recommended long-term, continuous monitoring of the NEO population and planetary-defense demonstration missions to be carried out for a decade at a cost of $250–$300 million annually, which would be one-sixtieth of the space agency's budget. Left unsaid, but clearly implied, was the fact that given the stakes, it would be a pittance.

- Acquire essential search, tracking, and warning capabilities for the early detection of potential NEO impactors and for tracking them with adequate precision with a space-based infrared telescope and investigate the development of low-cost, short-term impact warning systems.

- Investigate the nature of the impact threat, and specifically the physical characteristics of NEOs that most directly relate to planetary defense; that is, knowing the enemy in order to be able to stop it. That could entail deploying an infrared telescope in a Venus-like orbit, which would, in effect, allow the defensive unit to survey Earth's immediate neighborhood at a distance to provide as much warning time as possible of an impactor.

- Prepare an adequate response to the range of potential

impact scenarios, including testing innovative deflection technologies in space, assist agencies that are responsible for civil defense and disaster response, research ways to stop NEOs with nuclear and other weapons, and deflect them.

- Provide leadership for the government to address planetary-defense issues with other agencies, in public education, in the news media, and in international forums; to support research in the physical, environmental, and social consequences of a range of attack scenarios; and, with other relevant agencies, to develop a planetary-defense communication plan.

David Morrison heartily agrees. "For the non-science policy maker, the impact hazard is a complex problem featuring the interactions of physical, technical, and social systems under conditions of great uncertainty," he has written. "Communications are key, since in the end it is society's perception and evaluation of the hazard that are likely to determine what social and economic resources are applied. Policy makers will be dealing implicitly with the costs of action vs. the costs of inaction. From their perspective, even such an 'innocent' first step as the Spaceguard Survey may have substantial social or political costs—for example if frequent 'false alarms' persuade the public that scientists are incompetent and are squandering public funds, or if the existence of a survey triggers public demand for more expensive defense systems that decision makers are not prepared to provide."[36]

While the first small steps are being taken by space agencies to stop dangerous asteroids and comets, Morrison has explained, "we are a long way from the technology to deflect an asteroid, especially one potentially threatening to civilization. However, it seems reasonable to expect that if such a large asteroid is discovered, one whose impact could kill a billion people, the spacefaring nations would find a way to deflect it and save the planet. Given such a specific threat, almost any level of expense could be justified. This effort would represent the largest and most important technological challenge ever faced; whether it met with success or failure, world civilization would be forever changed."[37]

Given what is at stake, the ultimate Strategic Defense Initiative to defeat impactors as an integral component of an international planetary-defense system should therefore be a top priority for the world community and would obviously be worth any level of expense.

That system has been seriously pondered. It is called NEOShield, an excellent name for a peer-reviewed, collaborative, international research program that was started in January 2012 by a handful of science organizations, including the French National Center for Scientific Research and universities in France, Germany, Great Britain, the United States, Spain, and Russia. It is funded in large part by the European Commission, which is the executive body of the twenty-eight-member European Union, a regional unification organization, and is coordinated and essentially led by DLR, the German national space agency. The idea is to study various ways to detect and prevent an asteroid or comet from connecting with Earth and to decide which are the most workable. The research includes laboratory tests and sophisticated computer modeling of what an NEO would do on approach and what techniques would be best for deflecting it. The Boy Scouts famous motto is appropriate: Be Prepared. The participants would like to have a defensive plan in place, should the need arise, rather than have to frantically come up with something quickly as the emergency worsens.

The three most promising defensive measures seem to be the use of so-called gravity tractors to nudge a potential impactor off course, a kinetic shot to smash it, and explosive blast deflection. (In the most honest of all possible worlds, the last would also be called the Teller Technique, since the explosive blast would have to be nuclear.) The commission's concern about meteorites striking its member states is well founded. People all over Europe, like people everywhere else, have been seeing meteorites and, closer still, meteoroids, streak across the sky since they lived in caves. And in almost every European country, they have collected pieces of the space rocks that were seen to explode into fragments and rain down on their communities. The proof, at least for them, is in the touching.

"NEOShield was launched in mid-January and, over the next three and a half years, will investigate the measures that can be employed to prevent near-Earth objects such as asteroids and comets from colliding with Earth," DLR officials have said. Alan Harris, a senior scientist and NEOShield project leader at the German Aerospace Center's Institute of Planetary Research in Berlin, added "The scientific side of this will include the analysis of observational data on NEOs and laboratory experiments in which projectiles are fired at asteroid surface analog materials with different compositions, densities, porosities and structures. We'll be looking in detail at the tricky technical issues associated with autonomous control of a spacecraft in the immediate vicinity of a large, potato-shaped asteroid, and ion thrusters that may have to function continuously and reliably over a period of 10 years or more," Harris explained.[38]

Spotting a potential impactor very far out—at least twenty-five years is most often mentioned—and then pushing it ever so slightly off course so that it misses Earth by a substantial margin is the perfect, optimal strategy, and it is feasible if the asteroid is very far away. But it obviously will not work in a surprise attack, and it is surprise attacks that have marked this planet's experience with NEOs. Every airburst and direct hit has come without warning. But no effective defense can be mounted without a very long warning time. A tracking system with antennas stationed roughly 120 degrees apart around the planet and looking out in all directions, plus Sentinel, should therefore be set up to provide enough warning time so that an effective defensive operation can be undertaken in time to be effective.

The defensive operation has a theoretical precedent of sorts, though it was for a threat from elsewhere on the planet, not from space. It was the Strategic Defense Initiative—SDI—or Star Wars, as it was called by its many opponents (this author having been an early one)—that was inspired by Edward Teller and that led to President Reagan's famous Star Wars speech on March 23, 1983, and the antiballistic-missile-design stampede that followed for billions of defense dollars. The plan called for Soviet ballistic missiles to be bludgeoned from space as they rose out of their silos and then attacked continuously in their midcourse and terminal phases.[39]

Scientists and engineers at some of the national laboratories and in the corporate sector went into a virtual feeding frenzy to come up with concepts that would bring government contracts that Congress set at $44 billion between 1983 and 1993. Lowell Wood invented a nuclear-pumped x-ray laser that would have, in theory, stopped ballistic missiles in midflight by zapping them with very powerful laser beams that came out of hydrogen-bomb explosions. Theory, however, did not translate to fact. The laser did not work. But it was only one among several space weapons that were concocted to stop incoming warheads. The hypervelocity railgun, which was technically the Compact High Energy Capacitor Module Advanced Technology Experiment, or CHECMATE (another GAD award contender), was basically pellets on a rail that were accelerated by an electrical charge to such a high speed that they could drive a bullet through a tank's armor. And researchers devised particle-beam weapons that used high-energy beams of atomic or subatomic particles to destroy a target by damaging its atomic and molecular structure; in effect disassembling it by breaking it up on its most fundamental level. Besides the x-ray laser, there was a chemical laser, space-based interceptors, and even something called Brilliant Pebbles, which were watermelon-sized projectiles that were to be made of tungsten and fired at the missiles from above the way pellets are fired from a shotgun. The missiles were to be spotted and tracked by sensors called Brilliant Eyes. The Strategic Defense Initiative was a profoundly bad idea not only because it was destabilizing rather than stabilizing, but also because the spacecraft that were in permanent orbit and that were supposed to fire the lasers and launch the antimissile missiles were themselves vulnerable to an attack by the Soviets prior to the missile fusillade fired at the United States. John Pike of the Federation of American Scientists correctly called SDI a "playpen for engineers."

SDI's problem was political not technological. The concept of creating high-tech space weapons to stop a missile attack from elsewhere on Earth was innovative, and it was only infeasible because it was directed against another superpower that could have rained so many nuclear warheads on the United States that an unacceptable number would have hit their targets, destroying not only strategic military installations but also many cities and killing millions. Unless they are directed by intelligent extraterrestrials, that quandary does not exist with NEOs, so a planetary Strategic Defense Initiative using ultra-high-tech weapons modeled on those that were conceived for SDI is not only feasible but imperative. The first-ditch defense should consist of easing a threatening object off course decades before impact with a gravity tractor. Given that distance, nudging it even a couple of centimeters so far out would be sufficient to cause it to miss Earth by a comfortably wide margin. But it is imperative that a last-ditch defense also be in place to stop any impactor larger than 140 meters that is on a collision course with this planet.

Scientists at California Polytechnic State University and the University of California, Santa Barbara, think so too and have therefore come up with DE-STAR, the Directed Energy Solar Targeting of Asteroids and exploration (yet another contender . . .), which would convert solar energy into a laser blast that would obliterate any large rocks or icicles bearing down on Earth. "The system is not some far-out idea from *Star Trek*," Gary Hughes, a professor and researcher at Cal Poly said. "All the components of the system pretty much exist today. Maybe not quite at the scale we'd need—scaling up would be the challenge—but the basic elements are all there and ready to go. We just need to put them into a larger system to be effective, and once the system is there, it can do so many things." Philip Lubin, a professor of physics at the University of California, Santa Barbara, added that "our proposal assumes a combination of baseline technology—where we are today—and where we almost certainly will be in the future, without asking for miracles."[40]

There is indeed a consensus that miracles will not be necessary, a point that echoes Gene Shoemaker's observation about the difference between earthquakes, hurricanes, volcanoes, and other natural disasters that are generated on this planet and objects that come from elsewhere. The former cannot be controlled (at least not yet). The latter can.

There is complete agreement among the most sagacious that planetary defense starts with knowing precisely what is out there and where it is headed. As would be expected, the master tome on the subject, *Hazards Due to Comets and Asteroids*, was published by the University of Arizona Press in 1994. One of its chapters, written by David Morrison, Clark Chapman, and Paul Slovic (of the University of Oregon), described the impact hazard in considerable detail and concluded, as do their colleagues, that the largest "projectiles" have to be dealt with first and that the means to do so requires a comprehensive survey of Earth-crossers such as has been accomplished by the Spaceguard Survey. "Better understanding of the numbers, orbital distributions, and physical properties of asteroids and comets are required in order to define an effective defense system," they note.[41]

Then there was some significant progress in literally sizing up the problem. In August 2002, NASA was so pleased at the progress that was being made by the Spaceguard Survey to find and catalog 90 percent of Near-Earth Objects a kilometer or larger in diameter that it chartered a Science Definition Team to study the feasibility of extending the search for NEOs to those smaller than a kilometer. The team, whose twelve notable members included Yeomans, Robert S. McMillan, and Simon P. Worden, deliberated for a year and issued an important 154-page report in August 2003 that made three recommendations: (1) goals related to the search for potential impactors should be stated explicitly in terms of the statistical risk that would be eliminated and should be based on the standard cost/benefit analysis, (2) an NEO search program should be started that would discover and catalog the potentially hazardous "population" so precisely that 90 percent of objects smaller than the kilometer threshold (i.e., Earth threatening) would be eliminated, and (3) an announcement of opportunity should be released that would allow any individual or organization interested in developing the discovery and cataloging program to make specific recommendations.[42] That is Sentinel's assignment. Where planetary defense is concerned, it is literally a giant step in the right direction.

THE SURVIVAL
IMPERATIVE

"The Earth is so small and so fragile and such a precious little spot in the universe that you can block it out with your thumb," Rusty Schweickart has said. "And you realize on that small spot, that little blue and white thing, is everything that means anything to you—all of history and music and poetry and art and death and birth and love, tears, joy, games, all of it on that little spot out there that you can cover with your thumb. And you realize from that perspective that you've changed, that there's something new out there, that the relationship is no longer what it was."

Gene Cernan echoes that. "You . . . say to yourself, That's humanity, love, feeling and thought. You don't see the barriers of color and religion and politics that divide this world. You wonder if you could get everyone in the world up there, would they have a different feeling?"[1]

Their view of Earth as a solitary island of precious life in a dark and foreboding universe, an incubator of creatures with infinitely complex intellects and emotions, is shared by everyone who has gone to the Moon and by space scientists like Carl Sagan, who characterized it as a pale-blue dot to be cherished and protected.

That's not the way Osepok Tarov sees it, though. She is the commander of a multigenerational intergalactic spaceship that has left Earth and is heading for some place very far away to start a colony in Buzz Aldrin and John Barnes's epic science fiction novel *Encounter with Tiber*. Earth, she has decided, has become so spoiled that it is no longer a fit habitat for humans. "They found forest fire ashes, evidence of earthquake collapses, layers of volcanic ash, all kinds of things in the area," Osepok tells her crew to justify leaving the home planet. "We might have started here, but it was a tough place to stay alive. So we spread out—down the rivers, up into the hills, across the plains, eventually over the seas to Shulath—and now out into space. There's not a place in the universe that's safe forever; the universe is telling us, 'Spread out, or wait around and die.'"[2]

When Osepok thought about the danger of living on Earth, though, it was not just forest fires, earthquakes, and volcanoes she had in mind. It was The Intruders and the "bombardments" they caused. "The Intruder shattered into billions of pieces of all sizes, scattering into a great cloud. Thus, although the dense central part of the cloud missed our world by a wide margin, the debris—abundant even in the thin edges of the cloud, the biggest pieces the size of mountains, most boulder-sized or smaller— had sprayed our world, and Sosahy, in what the history books called the First Bombardment. The First Bombardment had been bad enough; one out of eight people worldwide killed, and Shulath wrecked. The Second Bombardment would finish off Nisu. One hundred and forty-some years in the future, there would be nothing left of us—unless some of us, somehow, could be somewhere else."[3]

Robert Shapiro, a brilliantly imaginative professor of chemistry at New York University, was not a fatalist about Earth, but he also believed that it is prudent to spread out. That is why he conceived of the idea of maintaining a continuously updated record of the home planet's civilizations at either pole and on the Moon so that if a catastrophe occurs, what happened to all the papyruses when the Great Library at Alexandria was destroyed will not happen again. "No skipper goes to sea thinking his boat is going to sink," Shapiro told the author, "but he carries a dinghy, life preservers and insurance just in case."[4]

The Moon is Earth's dingy—a seaworthy habitat that can help assure the survival of life if the mother ship is attacked or founders—but it is rarely seen that way. It has traditionally been considered to be the first stop in the exploration of the Solar System and beyond in what would be the greatest adventure in history, the first off-planet staging area in a great migration to space by a race of creatures that is genetically programmed to explore, to seek new worlds, in the tradition of Magellan, Columbus, Cortez, da Gama, Cheng Ho, and ibn Battuta.

That is what Jules Verne had in mind when he penned *From the Earth to the Moon* and its sequel, *Around the Moon*. The latter of the two was published in 1870 and, in it, three members of the Baltimore Gun Club—Barbicane, Ardan, and Nicholl—are shot out of the giant Columbiad cannon to scout the Moon for signs of a past civilization and to reconnoiter it for future habitability. (The cannon was necessary because it would be thirty-three years before Konstantin Eduardovich Tsiolkovsky, a rural Russian school teacher, published the seminal "Exploring Space with Reactive Devices" in the *Scientific Review*, which established liquid rockets as a viable propulsion system. Ironically, he was inspired to invent the liquid-propelled rocket, which developed far more thrust than the Chinese type that ran on powder like their famous fireworks, by reading Verne.)

Sure enough, the three adventurers see signs of an ancient civilization as they pass over the lunar surface, and Barbicane becomes convinced that humans could, with great difficulty, make the Moon habitable again.

"My friends," said Barbicane, "I did not undertake this journey in order to form an opinion on the past habitability of our satellite; but I will add that our personal observations only confirm me in this opinion. I believe, indeed I affirm, that the moon has been inhabited by a human race organized like our own; that she has produced animals anatomically formed like the terrestrial animals: but I add that these races, human and animal, have had their day, and are now forever extinct!"[5]

The master creator of supreme adventures then has one of the characters predict that the same fate could be in store for Earth. "And so," asked Michel Ardan, "humanity has disappeared from the moon?"

"Yes," replied Barbicane, "after having doubtless remained persistently for millions of centuries; by degrees, the atmosphere becoming rarified, the disc became uninhabitable, as the terrestrial globe will one day become by cooling."

"By cooling?"

"Certainly," replied Barbicane; "as the internal fires became extinguished, and the incandescent matter concentrated itself, the lunar crust cooled. By degrees, the consequences of these phenomena showed themselves in the disappearance of organized beings, and by the disappearance of vegetation. Soon, the atmosphere was rarified, probably withdrawn by terrestrial attraction; then aerial departure of respirable air, and the disappearance of water by means of evaporation. At this period the moon, becoming uninhabitable, was no longer inhabited. It was a dead world, such as we see it today."

"And you say that the same fate is in store for the earth?"

"Most probably."[6]

How Earth succumbs to the destructive universe that engulfs it is less important than the fact that Verne addressed its vulnerability and also its relationship with its solitary companion, which circles close by. The home planet and its moon have always had a special relationship. Indeed, the fact that the Moon is as large as it is relative to Earth and goes through predictable phases has made it the subject of veneration and worship throughout history. The oldest written records from ancient Egypt, Babylonia, India, and China tell of disparate societies that were convinced that the Moon had supreme power. It has been and still is worshipped because its phases have been associated with life and death: with the growth and decline of plant, animal, and human life, all of which are intimately related and mutually dependent. In Chinese mythology, the goddess Chang'e is stranded on the Moon after committing the sin and drinking a double dose of an immortality potion.

The Moon's supreme power has manifested itself in evil, too, which is why it has been venerated by witches since medieval times. The Moon's size, closeness, composition, and effect on Earth have led some to think of it not as a moon at all, which connotes its being an attendant body, but as another, smaller planet; that the Earth-Moon relationship is in fact that of a double planet. That carries the clear implication that humans and other creatures could live on it.

In Edward Everett Hale's opinion, they could also live on a moon that is made of bricks. The author of the short story "The Man without a Country" also wrote "The Brick Moon" as a three-part serial that was published in the *Atlantic Monthly* in October, November, and December 1869. It and was followed by "Life in the Brick Moon" in February 1870. They are serious adventure stories about people living in a manmade structure, and, understandably, they contain both errors in science and some excellent insights. But they are far more important as prophesy than literature because they describe people living harmoniously and productively in a space station that was made of bricks and was a precursor to one that is made of metal and called the International Space Station. It established the notion in nineteenth-century America that humans could fare well living in a structure that was not on Earth and, in doing that, it set the basis for an idea that would germinate into the concept of self-contained communities in space, including on the Moon.

Arthur C. Clarke was all in favor of establishing a permanent base on the Moon, not as a refuge from some calamity on Earth, but as the logical first step in humankind's inevitable expansion into space that he thought should include establishing colonies on Mars and beyond. He believed without question that it is humanity's destiny to spread throughout the Solar System and adapt to new environments as necessary. "Only little minds are impressed by size and number. The importance of planetary colonization will lie in the variety and diversity of culture which it will make possible—cultures as different in some respects as those of the Eskimos and the Pacific Islanders. They will, of course, have one thing in common, for they will all be based on a very advanced technology. Yet though the interior of a colony on Pluto might be just like that of one on Mercury, the different external environments would inevitably shape the minds and outlooks of the inhabitants. It will be fascinating to see what effects this will have on human character, thought and artistic creativeness," he wrote in *The Exploration of Space*, which was published in 1951, the third of about a hundred or so books he would write in his extraordinarily productive career.[7]

And, logically, the exploration was to start on the Moon, which he declared should be a self-sustaining colony to be developed as a thriving, productive enterprise—a lunar base, as he put it—not as a refuge. The idea, though he would not have put it this way, was to homestead the Moon and then extend the homesteading ever deeper into space the way Americans did when they moved west in wagon trains in the nineteenth century. (The size, safety, and creature comforts of spaceships being what they undoubtedly will be, trains of them will not be necessary unless aliens—"greenskins," as they might be called in politically incorrect space jargon—are encountered wearing feathers and are armed with bows that shoot laser arrows. Then the spaceships will have to be formed in circles with the Earthlings making certain to shoot their lasers to the outside . . .) "Today we can no more predict what use mankind may make of the Moon than could Columbus have imagined the future of the continent he had discovered. . . . Nevertheless, it is possible to foresee certain lines of development which appear likely as soon as we have reached the Moon, and we can also discuss, in general terms, the problem of making it habitable," Clarke continued.[8]

The problem, understandably, was to make the Moon like Earth, in that it would have air, water, vegetation, and all else that is necessary to sustain human life. A full-page color illustration in *The Exploration of Space* shows the lunar base consisting of four domed structures, one of them quite large, built into a rock formation with a large communication antenna close by and several others off in the distance, cultivated land, excavation underway, a rocket being launched in the distance, and a small truck carrying something in the foreground. The point was obviously to show the lunar base as a thriving metropolis that gets its energy from a solar power plant, a source of clean and relatively inexpensive energy that would also have a place on Earth. Yet Clarke also mentioned a subterranean enclave. "There is a good deal to be said for moving the lunar base underground at the earliest opportunity, if it proves possible to excavate the moon's surface rocks without too much difficulty. An underground settlement would be easy to air- and temperature-condition, and its construction would not involve carrying materials from Earth. Possibly suitable caves or clefts might be found which could be adapted."[9] The Martian base is set in a green countryside with vehicles on roads, looks like Philadelphia or Detroit, and is set under a colossal Plexiglas dome.

Writing when he did, Clarke had no way of knowing that getting to the Moon was going to be motivated by Cold War politics, not by the desire to invigorate the human spirit, to mine the lunar landscape for precious metals to supplement those on Earth, or to spread out for safety's sake. The Soviet Union had tried, without success at first, to send probes to Mars and Venus and dispatched robotic Luna spacecraft to the Moon in evident preparation to send cosmonauts there, so President John

F. Kennedy's advisers convinced him that landing astronauts on the Moon would be a feat of supreme and lasting importance because it would prove conclusively to the world in the most dramatic and unchallengeable way that the United States could land Americans on another world and get them safely home again because its system was economically, technologically, and politically superior to that of a totalitarian union of socialist republics. And it worked. The landing of the *Eagle* on the Sea of Tranquility on July 20, 1969, was a historic triumph of immense proportion. But it was done for bragging rights, for expediency, not for the long-term benefit of civilization in any capacity.

Yet Clarke was a scientist, not a politician, so he saw the Moon as both a scientific asset and one that contained abundant natural resources that could help nourish Earth. "Within a few years of the first landing, it should be possible to establish a small camp on a permanent basis, keeping it supplied by a regular service of ships from Earth. A great effort would be made to set up an observatory with a telescope of moderate size: in fact it would be worthwhile building a spaceship for no other purpose than to carry a reflecting telescope of, say, twenty-inch aperture to the Moon. . . . The Moon has so many advantages as an observatory that future generations may well wonder how we discovered anything about the heavens while we were still 'earthbound.'"[10]

An observatory on the Moon could be used to find and track asteroids, a fact that Clarke did not address. In common with many other scientists and others in the space community, he thought that asteroids are loaded with natural resources that can be exploited to help Earth, and he minimized the potential danger they pose. Former astronaut Tom Jones, a veteran of four space-shuttle missions and whose memoir, *Sky Walking*, predicts that near-Earth asteroids (NEAs) offer advantages for commercialization as well as exploration of NEAs, agrees. "The water and other mineral resources that we know are present on some NEAs could help reduce the long-term costs of exploring the Moon, Mars, and the rest of the solar system. And in the course of exploring them, we can test the technology needed to divert any asteroid on a collision course with Earth. . . . From my spacefarer's perspective," he continued, "the most attractive idea about 'astronauts to asteroids' is that such voyages represent a natural progression in difficulty, more challenging than the dash-for-the-Moon Apollo missions but less daunting than the multiyear duration of a Mars landing expedition."[11]

John W. Young is concerned about the asteroid threat and about humanity's survival in general, which is why he has become a strong advocate for colonizing the Moon, a place he knows firsthand. In 1981, he commanded the space shuttle *Columbia* on the first flight that orbited Earth; in total, he spent forty-two years with NASA, during which he became very familiar with the asteroid situation. He also became intimately familiar with the Moon, since he has been there twice. He orbited it during the Apollo 10 mission in late May 1969 in a dress rehearsal for the Apollo 11 landing in July, and he was the ninth person to walk on the lunar surface when he commanded the Apollo 16 mission in April 1972 and spent seventy-one hours there, including a romp with Charles Duke in a lunar roving vehicle.

"I started advocating [colonizing the Moon], so as to enable humankind to survive even if Earth gets hit hard by an asteroid, we must continue exploring the Solar System," he wrote in *Forever Young*, his autobiography. "Specifically, we need to build a permanent human base on the Moon where people from different nations can live and work. If we can learn how to terra-form on the Moon, the same technology could save Earth inhabitants from the long nuclear winter that would be caused by an asteroid impact." The reference to a nuclear winter, in which debris would cloud the sky for months, blocking sunlight, implies that the impactor he had in mind was quite a bit larger than the one that grazed Chelyabinsk.

Young is respectful of NASA for much of what it has accomplished, certainly including the Apollo program and Solar System exploration, but he nonetheless faults it for not making a strong case for a return to the Moon, not for glory, but for the safety of the world in the broadest sense: planetary defense:

> In trying to persuade the public why we need to go back to do more human exploration of the Moon, has NASA chosen to explain that such exploration will provide us with much of the advanced technologies that are badly needed to ensure the long-term survival of our threatened and endangered species on Earth? No. Has NASA made a powerful enough case that the Moon is the very best place to establish the first human bases for living, working and supporting Earth's people in this, the twenty-first century? No. It's no doubt because NASA's bureaucracy sees no political advantage in scaring people. But I see it differently. The human race is at war. Our biggest enemy, pure and simple, is ignorance. The bottom line of all human exploration is to preserve our species over the long haul. We have no idea how much time we have left. The Solar System and Planet Earth are talking to us. But no one is listening. There are major events that can "do in" our civilization. And in time they most certainly will.[12]

Clarke not only envisioned the Moon as an observatory that would not be impeded by an atmosphere—and therefore a superior vantage point to search for threatening asteroids, though he did not mention that—but, far more important, he saw it as the logical first place for people and other creatures to settle and homestead as they migrated inevitably toward Mars and the other outer planets. Although he probably would not have put it this way, for Clarke, the need to explore and migrate to worlds beyond this one—wanting to see what is beyond the horizon and venture there—is as embedded in the human genome as romantic love, protection of family, and seeking safety from the extremes of nature. The last is what impelled Osepok to get out of town, and it, together with the migration impulse, is probably what got Clarke to decide that colonizing the Moon would be the logical first step in the great expansion to other worlds.

> The greatest technical achievements of the next few centuries may well be in the field of what could be called "planetary engineering"— the reshaping of other worlds to suit human needs. Given power and knowledge (wisdom is rather useful, too) nothing that does not infringe on the laws of Nature need be regarded as impossible. We will return to this theme when we discuss the other planets, but it will already be apparent that the conquest of the Moon will be the necessary and inevitable prelude to remoter and still more ambitious projects. Upon our own satellite with Earth close at hand to help, we will learn the skills and techniques which may one day bring life to worlds as far apart as Mercury and Pluto.[13]

Scientists will be among the first to say that social generalizations are dangerous. But they will also say, without stigmatizing biology or chemistry in the slightest, that physicists are known to be the most imaginative of the breed. Clarke was primarily a physicist, though his degree was in both physics and chemistry. Hans Bethe was a physicist, and so were Theodore Taylor, Edward Teller, and the exuberant and irrepressible Richard Feynman, a theoretical physicist from one of New York City's earthier neighborhoods, who taught at Caltech (including freshmen) and proudly boasted that he was first and foremost an intellectual "explorer," knowing that exploration in its many forms contributes so much to the glory and nobility of the human spirit.

To that extraordinarily imaginative group of notables, add Gerard K. O'Neill, who was a professor of physics at Princeton University, and a prophet who decided that spreading out by creating a self-sustaining colony in space was unarguably imperative, given Earth's finite capacity to nourish the creatures that live off it and the multiple dangers out there. He got that out in a landmark work, "The Colonization of Space," which appeared in *Physics Today* in 1974 and was expanded into the book *The High Frontier: Human Colonies in Space*, which was published two years later and immediately became the bible of the migration to space movement. (Osepok undoubtedly would have loved it, even though her reasons for abandoning Earth were cynical, not curious and adventurous in the tradition of exploration.)

"How can colonization take place?" O'Neill asked, rhetorically in "The Colonization of Space." He continued:

> It is possible even with existing technology, if done in the most efficient ways. New methods are needed, but none goes beyond the range of present-day knowledge. The challenge is to bring the goal of space colonization into economic feasibility now, and the key is to treat the region beyond Earth not as a void but as a culture medium, rich in matter and energy. To live normally, people need energy, air, water, land and gravity. In space, solar energy is dependable and convenient to use; the Moon and asteroid belt can supply the needed materials, and rotational acceleration can substitute for earth's gravity.
>
> Space exploration so far, like Antarctic exploration before it, has consisted of short-term scientific expeditions, wholly dependent for survival on supplies brought from home. If, in contrast, we use the matter and energy available in space to colonize and build, we can achieve great productivity of food and material goods. Then, in a time short enough to be useful, the exponential growth of colonies can reach the point at which the colonies can be of great benefit to the entire human race.[14]

What O'Neill had in mind for the human habitat was a pair of cylinders that would be between sixteen and twenty miles long and four miles in diameter. They would be self-contained worlds that are copies of the most attractive and hospitable parts of the home planet, including dwellings, parklands and forests, lakes, rivers, grass, trees, animals, and even birds (but not farmland; agriculture would take place somewhere else).[15] "Birds and animal species that are endangered on Earth by agricultural and industrial chemical residues may find havens for growth in the space colonies, where insecticides are unnecessary, agricultural areas are physically separate from living areas, and industry has unlimited energy for recycling." He emphasized that each space colony would be a complete, self-contained ecosystem that could thrive independently of Earth.

"With an abundance of food and clean electrical energy, controlled climates and temperate weather, living conditions in the colonies should be much more pleasant than in most places on Earth," he continued. "For the 20-mile distances of the cylinder interiors, bicycles and low speed electric vehicles are adequate. Fuel-burning cars, powered aircraft and combustion heating are not needed; therefore, no smog. For external travel, the simplicity of engineless, pilotless vehicles probably means that individuals and families will be easily able to afford private space vehicles for low-cost travel to far distant communities with diverse cultures and languages."

And while living away from the home planet for a lifetime has usually been considered anathema to most people, O'Neill saw a benefit to it. "The self-sufficiency of space communities probably has a strong effect on government. A community of 200,000 people, eager to preserve its own culture and language, can even choose to remain largely isolated. Free, diverse social experimentation could thrive in such a protected, self-sufficient environment." He also noted that the communities would be protected from cosmic rays by the depth of the atmosphere and by land and steel supporting structures, but he erred in claiming that meteoroid damage should not be a serious danger because of their composition. "Most meteoroids are of cometary rather than asteroidal origin and are dust conglomerates, possibly bound by frozen gases; a typical meteoroid is more like a snow-ball than like a rock." Comets, as has been noted, are indeed more like snowballs than rocks, but meteoroids are small rocks that are derived from meteors and, hence, asteroids, and they are therefore also rocks, not chunks of ice.[16]

Gerard O'Neill was not the first to envision a large, self-sustaining colony in space. But he was a modern visionary whose detailed description of one and what would be necessary to keep it functioning so captivated Carolyn and Keith Henson that they created the L5 Society in Tucson in 1975, the year after "The Colonization of Space" was published, to carry his dream to fulfillment. The group took its name from the Lagrangian points, where the gravitational pull from Earth and from the Moon cancel each other out, so the colony would stay put without drifting off.

But there was an obstacle. It was called the Agreement Governing the Activities of States on the Moon and Other Celestial Bodies, or the Moon Treaty for short, which turned over the jurisdiction of all celestial bodies to the international community in a "common heritage of all mankind" morass. The Moon Treaty was born out of a palpable fear after Americans landed on the Moon that either side, but most probably the Americans, would build a missile base there with the weapons trained on targets on Earth. Missile defense was hard enough when they came from the other side of the North Pole, but it would be next to impossible to stop a barrage of them raining down from lunar launchers. And, beyond that, there was a feeling that a nation that controlled the Moon would have some unarticulated but real advantage over the other nations on this planet, most likely having to do with its vast natural resources. The treaty therefore held the United States to the spirit of Neil Armstrong's famous declaration about he, Aldrin, and Collins being there for all humankind.

So the agreement stipulated that the Moon had to be used solely for the benefit of all nations and peoples, and activities on it were therefore subject to international law. That put it off-limits as the site of a farm for the space colony. More important, the treaty prohibited any form of sovereignty or private property in space, which would have been the death knell for the colony itself. Nations with space programs—Russia, China, Japan, and India—did not ratify the treaty, and the L5 Society saw to it that the United States did not ratify it either. In the end, however, the colony literally died of its own weight. Calculations showed that the cost of rocketing the many thousands of tons of material to the Lagrangian points where Island One, as it was also called, would be constructed would be $96 billion.

Freeman Dyson, another theoretical physicist (though from Princeton, not Caltech), thinks that that sum is "preposterously large" to spend on a single enterprise, even on Island One.[17] In his autobiography, *Disturbing the Universe*, which celebrates science's contribution to civilization, he notes that O'Neill claimed that the $96 billion would be repaid within twenty-four years and asserts that, in reality, the space colonists would have to work for 1,500 years to pay each family's share of that staggering debt. The answer for Dyson is governmental. "It must inevitably be a government project, with bureaucratic management, with national prestige at stake, and with occupational health and safety regulations rigidly enforced. As soon as our government takes responsibility for such a project, any serious risk of failure or of loss of life becomes politically unacceptable. The costs of Island One become high for the same reason that the costs of the Apollo expedition were high. The government can afford to waste money but it cannot afford to be responsible for a disaster."[18]

William Sims Bainbridge, a sociologist who is the first Senior Fellow at the Institute for Ethics and Emerging Cultures and was on the faculty of Towson State University, had a breakthrough more than two decades ago when he predicted that the colonization of space would not only preserve humanity but also profoundly change it by diversification. "Among the potentially radical results of space flight is the utter transformation of human society," he wrote in *Goals in Space: American Values and the Future of Technology*, published in 1991.

Beyond the pull of Earth's gravity we could create "new cultures," thus achieving "increased cultural diversity." Establishing "societies on other planets" would lead "to hitherto unknown life-styles," "creating an alternate pattern of life that a lot of people will enjoy." This is to be expected from the space program partly because "knowledge and technology change life-styles and beliefs over time." Such changes may be essential "to develop alternatives to current ways of life that are careening towards nuclear destruction."[19]

Bainbridge went on to explain that the material condition of colonies would differ, requiring different modes of life, and that social isolation would permit "cultural drift and random innovation." And he quoted Dyson very effectively: "It is in the long run essential to the growth of any new and high civilization that small groups of people can escape from their neighbors and from their governments, to go and live as they please in the wilderness. A truly isolated, small, and creative society will never again be possible on this planet."[20]

The term *conquest* connotes triumph over a dangerous opponent in a contest—an opponent that could be a wilderness—and that is the way humankind's relationship with space was often couched before the Space Age. *Outer space* (as it was called to emphasize its awesome vastness) was known to be airless, gravity-free, and subject to the kind of deadly radiation that does not reach Earth because of the atmosphere, so it was considered an alien environment that would have to be conquered like the Himalayas and other high places. That is why the place where spaceships from Earth landed on the Moon would be prospected by men wearing "spacesuits," Clarke explained, which would be rigid because of the pressure inside them "like an inflated tyre, thus spread-eagling the unfortunate occupant."[21] The *Collier's* series "Man Will Conquer Space Soon!" anticipated humankind's inevitable triumph over worlds beyond its own, including the "conquest" of the Moon, before the Space Age began with the launching of *Sputnik*.

Bruce C. Murray, a planetary geologist like Gene Shoemaker, who became prominent in Solar System exploration when he interpreted the discoveries of the early Mars missions and went on to head the Jet Propulsion Laboratory (JPL), was a firm believer that the exploration of the Solar System has to be an international effort, not a nationalistic one. He made that clear in 1989, during the last flushes of the Cold War, when he envisioned an expedition to Mars that was also applicable to the Moon.

> For the taking of the great step to Mars, there is ample time. Unlike Kennedy's Apollo counterpunch in response to the Soviet space successes of 1957–61, the next great human leap to a new world need not be a nationalistic space race. It need not stress human and material resources as intensely as did Apollo or compete as furiously with other national and global imperatives.
>
> Next time we can go to Mars together in a resounding victory of human intelligence and spirit over runaway technological and political change. The U.S.-Soviet rivalry for world leadership can evolve from unrelieved military confrontation to sophisticated competition to lead and facilitate international cooperation.[22]

Murray also believed that equating the exploration of space with that of Earth was a stretch. "In fact, there is little in human experience upon which to project extraterrestrial colonization and migration. Sometimes Columbus's voyages to the New World are cited as a historical precedent. But those European endeavors were motivated and financed by parent-country expectations of economic return. The pursuit of individual goals in the New World—scientific curiosity, religious and political freedom, adventure and personal fulfillment—was possible mainly because of economically motivated kings and parliaments, who remained safe at home awaiting their financial return."[23]

That return, with other motivating factors, bears on establishing a permanent presence in space, a factor that Harrison H. Schmitt, geologist, protégé of Gene Shoemaker, and the last man on the Moon as part of the *Apollo 17* crew, supports. Of all the resources that abound on the Moon, helium-3 is the most important, in Schmitt's opinion, as well as others who are knowledgeable on the subject. "The financial, environmental, and national security carrot for a Return to the Moon consists of access to low-cost lunar helium-3 fusion power," he says. "Helium-3 fusion represents an environmentally benign means of helping to meet an anticipated eight-fold or higher increase in energy demand by 2050. Not available in other than research quantities on Earth, this light isotope of ordinary helium reaches the Moon as a component of the solar wind, along with hydrogen, helium-4, carbon and nitrogen. Embedded continuously in the lunar dust over almost 4 billion years, concentrations have reached levels that can legitimately be considered on economic interest. Two square kilometers of large portions of the lunar surface, to a depth of 3 meters, contains 100 kg (220 pounds) of helium-3, i.e., more than enough to power a 1,000-megawatt (one-gigawatt) fusion power plant for a year." And, ever the futurist, Schmitt envisions helium-3 being used to power settlements in space as well as on Earth. "By-products of lunar helium-3 production will add significantly to future economic returns as customers for these products develop in space. No such by-products are known that would warrant their return to Earth; however, locations in Earth orbit, on Mars, and elsewhere in deep space constitute potential markets. The earliest available by-products include hydrogen, water, and compounds of nitrogen and carbon. Oxygen can be produced from lunar water. Finally, metallic elements, such as iron, titanium, aluminum and silicon, can be extracted from mineral and glass compounds in the lunar regolith (soil)."[24]

Schmitt is also convinced that the cheap fuel is also potentially useful for space settlement and the necessary asteroid defense that goes with it. "An ultimate benefit to humankind from a lunar resources and helium-3 fusion power initiative will be its inherent potential to catalyze human settlement away from Earth. First of all, the initiative needs permanent habitation on the Moon to minimize operational costs. In addition, the immense technology base arising from such an initiative, the space life-supporting by-products of helium-3 production on the Moon, and helium-3 fusion technology adapted to interplanetary rockets will enable broad scale use of space stations, the initial settlement of Mars, the diversion of asteroids from a collision course with Earth, and human travel elsewhere in the solar system and possibly into the galaxy."[25] Harrison Schmitt believes that space is there for our progressive expansion, and helium-3 is a resource that nature has provided to help us achieve that expansion. Like others who have been to space, he understands that it is an infinitely vast combination of extraordinary, exhilarating beauty and is spiritually necessary for our cultural growth and fulfillment while at the same time being so environmentally alien that it can easily be deadly. He is certainly mindful of the fact that the large, speeding rocks are one of the things that can make it perilous.

There are two rational responses to that danger: repel the attackers in the ultimate strategic-defense initiative, and disperse in case repelling them fails. But to disperse is not to flee. It is the positive, creative process of spreading out to expand the human presence as far as possible: to settle along new frontiers and shape environments wherever possible in which humans and their collective culture can spread, take root, multiply, and thrive.

Alan Wasser is for the occupation of space, too, and is one of the growing number of entrepreneurs who are convinced that space settlement for both economic gain and safety is supremely important. He also believes that NASA, as a government institution, is not equipped to oversee the creation of the settlements and that unbridled capitalism—making space profitable—is a vital imperative for humanity's expansion and survival. He has been the chairman of the National Space Society Executive Committee, has been active in other space organizations, and has written a series of informed articles that propose US land-claims-recognition legislation to foster private enterprise to colonize the Moon and Mars. It is called the Space Settlement Initiative, and here is what Wasser says about it, with the asteroid threat obviously in mind:

> The settlement of space would benefit all of humanity by opening a new frontier, energizing our society, providing room and resources for the growth of the human race without despoiling the Earth, and creating a lifeboat for humanity that could survive even a planet-wide catastrophe.
>
> Space development has almost stopped, primarily because no one has a sufficient reason to spend the billions of dollars needed to develop safe, reliable, affordable transport between the Earth and the Moon. Neither Congress nor the taxpayers wants the government stuck with that expense. Private venture capital will support such expensive and risky research and development ONLY if success could mean a multi-billion dollar profit. Today, there is no profit potential in developing space transport, but we have the power to change that. We have the power to create a "pot of gold" waiting on the Moon, to attract and reward whatever companies can be the first to assemble and risk enough capital and talent to establish an airline-like, Earth-Moon "space line" and lunar settlement.

"The Lunar Line" would be a perfect name for the enterprise.[26]

Wasser has anticipated that some will invoke the 1967 Outer Space Treaty, which has been signed by more than one hundred nations and forms the basis for international space law, to show that his plan runs counter to the agreement. But he staunchly—and correctly—maintains that such is not the case. Among other things, the treaty prohibits any nation from claiming the Moon or any other celestial body as its own; that they are the common heritage of all humankind. "But quite deliberately," he says, "the treaty says nothing against private property. Therefore, without claiming sovereignty, the U.S. could recognize land claims made by private companies, regardless of nationality, that establish human settlements on the Moon or Mars."[27]

Randii Wessen, an engineer and deputy manager of the Project Formulation Office at JPL, agrees. Being with JPL has gotten him intimately involved in Solar System exploration, and he knows that it has fallen off abruptly. He also obviously knows that Elon Musk's company SpaceX has taken over supplying the International Space Station, while the space shuttles are relegated to being tourist attractions—"Museums"—with *Enterprise* at the Intrepid Sea, Air and Space Museum in New York; *Discovery* at the National Air and Space Museum in Washington; *Atlantis* at the Kennedy Space Center in Florida; and *Endeavour* at the California Science Center in Los Angeles.

Wessen sees a clear parallel between the development of aviation and commercial spaceflight. "The first thing that happens is the military wants one—they'll find it themselves. Next, the U.S. government says this is critical to national security or economic competitiveness, so we need to make up a job for these guys to keep them in business. For airplanes, the government said, 'We'll have you deliver mail.' They didn't need this service, but they gave it to airline companies to keep them going. This is analogous to spacecraft today. The government is saying [to companies like SpaceX], 'We want you to resupply the space station.' That's where we are now. As this gets more routine, these private companies are going to say, 'If we put seats on this thing, we'll make a killing.' They did it with airplanes. You can see that staring today, with four or five different companies who have suborbital and orbital launch capability."[28]

Annalee Newitz, a veteran science writer, has done extensive research on humanity's precarious existence on Earth and has come to the conclusion that spreading out is not only prudent but is also a matter of survival. "Eventually we'll have to move beyond patrolling our planetary backyard and start laying the foundation for a true interplanetary civilization," she has written in *Scatter, Adapt, and Remember: How Humans Will Survive a Mass Extinction*, which takes the position that, while defending against a threat is obviously justified, a good contract with nature, like a good contract with other people, has an escape clause. "Asteroid defense and geoengineering will only take us so far. We need to scatter to outposts and cities on new worlds so that we're not entirely dependent on Earth for our survival—especially when life here is so precarious. Just one impact of 10 on the Torino scale could destroy every human habitat here on our home planet. As horrific as that sounds, we can survive it as a species if we have thriving cities on Mars, in space habitats, and elsewhere when the Big One hits. Just as Jewish communities managed to ensure their legacy by fleeing to new homes when they were in danger, so, too, can all of humanity."[29] According to the NEO watchers' best calculations, the "Big One" is not due back for another century or so. But, again, as Robert Arentz put it succinctly, "It's not a matter if; it's a matter of when."

Planetary defense, which is to say the defense of life on this planet, requires two courses of action, both of them necessarily international and depending on close coordination. The attackers must be stopped, literally at all costs, by whatever means are judged to be the most effective. That requires a graduated response that begins with trying to nudge the potential impactor off course twenty-five years or more before the collision. If it will not be nudged, or if there is a surprise attack with the impactor discovered when it is dangerously close, then weapons—missiles or other projectiles, lasers, or nuclear weapons—will have to be used and should be ready. The Department of Planetary Defense should develop the weapons in close coordination with the international community.

At the same time, humanity should spread out by starting settlements first on the Moon and then beyond it. The strategy should be inherently positive, which is to say to explore new worlds and settle where we can, to realize our full potential as a species that is an integral part of this Solar System. Exploration is fundamentally important for nourishing the spirit as well as for providing the basis for homesteading new worlds. Spreading out should not be undertaken as an alternative to extinction or suffering in an abominable environment, which drove Osepok away, but as a supremely rewarding adventure in its own right, spiritually, aesthetically, and physically. And asteroids can be an asset as well as a threat and are therefore wonderful, exciting, and potentially profitable places to explore and settle. Besides being dangerous when they get too close, they are an inherently valuable resource, which shows that Antoine de Saint-Exupéry's little prince had it right after all.

ACKNOWLEDGMENTS

First and foremost, my deep appreciation goes to David Morrison, an astrophysicist and senior scientist at the Astrobiology Institute at NASA's Ames Research Center, who is also a charter member of the small group of highly knowledgeable NEO stalwarts in the international science community. He has shown, by example, how important it is to know about those objects. Furthermore, he read parts of the manuscript and not only made useful suggestions on what material to include but also corrected many errors. There was an expression we had at the *New York Times* (in a previous life of mine) that is applicable to Dave: he pulled me off the cross. And, obviously, mistakes in what he did not read are mine to bear.

Robert Arentz of Ball Aerospace and Technologies Corp., who served with me on the National Research Council (NRC) panel, has become a friend, a supplier of all manner of material for the book, and a wizened interpreter of developments having to do with asteroids, comets, and other things that go bump in the day and night. He, too, read part of the manuscript for mistakes and made many helpful suggestions. I was and continue to be exceedingly lucky to have him in my corner.

The genesis of this book developed at three planetary-defense conferences held around the country that were conducted by the NRC of the National Academy of Sciences. My presence at those meetings was at the invitation of Dwayne A. Day of the NRC, who invited me to attend as the only nonscientist on the fourteen-member Survey and Detection Panel. Dwayne extended the invitation in the evident hope that I would create a work that captured the spirit and importance of the situation as it was shown by the scientists who shared their findings with us. Given how many first-rate science writers are out there, I was and remain deeply complemented by his having selected me. I hope this book and some other relevant projects justify his decision.

Mariel Bard—the Bard of Amherst—copyedited this book, and that entailed quite a bit more than juggling punctuation marks and indenting paragraphs. Her alert eye spotted several instances of repetition, and she raised scores of excellent questions about phrasing, sources, and statements that were confusing or misleading. She practiced the highest level of her craft and, in doing so, became an invaluable partner in the creation of *The Asteroid Threat*.

Sean Mulligan, one of the first and best members of what is now the graduate Science, Health and Environmental Reporting Program at New York University, and who is now a professional dynamo and a valued friend, has not only alerted me to news of important "visitors" in the neighborhood but also has been my computer adviser. I am very grateful on both scores.

Neil deGrasse Tyson, the effervescent, boundlessly energetic, and always overcommitted director of the Hayden Planetarium in New York, was very generous with his time in discussing the asteroid threat and the feasibility of using the Moon as an archive and establishing a colony there. The title of his autobiography, *The Sky Is Not the Limit*, speaks to his vision of where we ought to be headed as well as to his life.

Elizabeth Paul, a reference librarian at the Westport Public Library, tracked down and gave me reviews of *Deep Impact* and *Armageddon*, films that figure importantly in why asteroids and comets so fascinate the public. Her help is very much appreciated as well. Robert "Bob" Shapiro (a professor of chemistry at New York University who is no longer here), Steven Wolfe, and I formed the Alliance to Rescue Civilization to archive civilization's record, and, in the process, I got to thinking about planetary defense, which led to this book. Bob had an extraordinary imagination and was endlessly creative. So is Steve, who is also a space enthusiast and an imaginative writer, and who worked for California governor Jerry Brown, another active space proponent; the synergy was obvious and infectious.

NOTES

CHAPTER 1. CHELYABINSK: RUSSIAN ROULETTE

1. Mayak accident: "Chelyabinsk: The Most Contaminated Spot on the Planet," wentz.net/radiate/cheyla/ (accessed February 12, 2014).

2. Surayev's snub: "Medal Snub for Russian Cosmonaut Sparks 'Cosmic Scandal,'" *Space Daily*, September 7, 2010.

3. The anger in Chelyabinsk: "Meteor Fragments Spark 'Gold Rush' in Russia," *Fox News*, February 18, 2013.

4. The meteor's characteristics: Matt Smith, "Planet of Sound: Meteor Blast Resonated Around Earth," CNN, February 27, 2013.

5. The blast and shock wave: Ibid.

6. Children injured; windows destroyed: "Meteorite Hits Russian Urals: Fireball Explosion Wreaks Havoc, up to 1,200 Injured," RT.com, http://rt.com/news/meteorite-crash-urals-chelyabinsk-283/ (accessed February 12, 2014).

7. Students' and Frenn's experiences: Ellen Barry and Andrew E. Kramer, "Meteor Explodes, Injuring over 1,000 in Siberia," *New York Times*, February 16, 2013.

8. Ms. Borchininova's remark: Andrew E. Kramer, "After Assault from the Heavens, Russians Search for Clues and Count Blessings," *New York Times*, February 17, 2013.

9. Ms. Nikolayeva's account: Ibid.

10. Kuznetsov's account: Phil Black and Laura Smith-Spark, "Russia Starts Cleanup after Meteor Strike," CNN.com, February 19, 2013.

11. Weird occurrences and Popova: Andrew E. Kramer, "In Russia, Ruins and Property Spared by Meteor, Side by Side," *New York Times*, February 18, 2013.

12. Yurevich's statement: "Meteorite Hits Russian Urals."

13. Number of injured and the broken spine: Ibid. Russian sources later reported the number of those who requested medical assistance at about 1,500.

14. Damage to structures in Chelyabinsk: "Meteorite-Caused Emergency Situation Regime over in Chelyabinsk Region," *Russia beyond the Headlines* (published by *Rossiyskaya Gazeta*), Interfax, March 5, 2013.

15. What the rocks are called depends on their distance from Earth. Asteroids, which are also called *minor planets* or *planetoids*, are mainly in the Asteroid Belt between Mars and Jupiter, but some strays pass in the vicinity of Earth but are not very close. Meteors are asteroids or other objects that enter Earth's atmosphere and either burn up, vaporize, or explode in a fireball that is called a *bolide*. They are made of iron or stone and become meteorites if they survive the plunge through the atmosphere.

16. 2012 DA14 close-approach prediction: "JPL Small-Body Database Browser," NASA/JPL, http://ssd.jpl.nasa.gov/sbdb.cgi?sstr =2012DA14;cad=1#cad (accessed February 21, 2014).

17. Hail of bullets: *Asteroids: Deadly Impact*, National Geographic Video, 2003.

18. Tunguska explosion: Henry Fountain, "Meteor Is Not Siberia's First Brush with Objects Falling from Space," *New York Times*, February 15, 2013.

19. Bissell: John Williams, "A Flash in Russian Skies, as Inspiration for Fantasy," *New York Times*, February 16, 2013; see also Tom Bissell, "A Comet's Tale: On the Science of Apocalypse," *Harper's*, February 2003.

20. Tyson: Clyde Haberman, "To at Least One Earthling, Siberia Meteor Proved That Science Is Vital," *New York Times*, March 11, 2013.

21. The Tyson asteroid: Neil deGrasse Tyson, *The Sky Is Not the Limit*, 108. Asteroid 9930Billburrows was named after the author. More than twenty thousand asteroids have been named after people, places, and things, including 558 Carmen, 588 Achilles, 1566 Icarus, 1569 Evita, 1864 Daedalus, 1930 Lucifer, 2001 Einstein, 3153

Lincoln, 3154 Grant, 3155 Lee, 3767 DiMaggio, 3768 Monroe, 4342 Freud, 7934 Sinatra, 8749 Beatles, 327 Columbia, 736 Harvard, 4523 MIT, 9769 Nautilus, 9770 Discovery, and 9777 Enterprise.

22. Medvedev's and Rogozin's remarks: Peter Fowler, "PM Medvedev Says Russian Meteorite KEF-2013 Shows 'Entire Planet Vulnerable.'" Newsroom America, February 15, 2013, http://www .newsroomamerica.com/story/347222/pm_medvedev_says_russian _meteorite_kef-2013_shows_entire_planet_vulnerable_.html (accessed February 13, 2014); Howard Amos, "Meteorite Explosion over Chelyabinsk Injures Hundreds," *Guardian*, February 15, 2013, http://www.theguardian.com/world/2013/feb/15/hundreds-injured -meteorite-russian-city-chelyabinsk (accessed February 13, 2014).

23. Theories on the cause of the explosion: Fred Weir, "Was Chelyabinsk Meteor Actually a Meteor? Many Russians Don't Think So," *Christian Science Monitor*, February 22, 2013.

24. Hoyle and Wickramasinghe's theory: Fred Hoyle and N. C. Wickramasinghe, *Diseases from Space*, ch. 1.

25. Yeomans on life in comets: Donald K. Yeomans, *Comets*, 350.

26. The damage estimate and Pavlovsky: Will Englund, "After the Meteor, Russian Residents Prepare to Clean Up," WashingtonPost .com, February 16, 2013.

27. The leaders' response: Kirit Radia, "Russian Meteor: Chelya-binsk Cleaning Up after Meteor Blast," ABC News, *World News*, February 16, 2013.

28. Sasha Zarezina and Larisa Briyukova: Andrew E. Kramer, "Russians Wade into the Snow to Seek Treasure from the Sky," *New York Times*, February 19, 2013.

29. Selling meteorites: Sergei Loiko, "Rubles from Heaven: Russians Scoop up Meteorite Chunks for Sale," *Los Angeles Times*, February 19, 2013.

30. Ads from Yevgeny and others: Ibid.

31. Maxim: Ibid.

32. Cherkova: Ibid.

33. Meteor Disneyland: "'Meteor Disneyland'? Russian City Looks to Meteor Site's Future," herocomplex.latimes.com.

34. Cocktail: Meagan Kyla, "Kiss My Asteroid [Drinks with Jack]," February 22, 2013, http://929jackfm.com/kiss-my-asteroid -drinks-with-jack/ (accessed February 13, 2014).

CHAPTER 2. CHICKEN LITTLE WAS RIGHT

1. The prince and the asteroid: Saint-Exupery, *The Little Prince*, 77.

2. Comets as drama: Shapiro, *Origins: A Skeptic's Guide to the Creation of Life on Earth*, 235.

3. Life from a comet: Hoyle and Wickramasinghe, *Lifecloud: The Origin of Life in the Universe*, 134.

4. Diseases from comets: Hoyle and Wickramasinghe, *Diseases from Space*, 8–9.

5. Shoemaker, "They *are* bullets": *Asteroids: Deadly Impact*, National Geographic Video, 2003.

6. Vredefort crater: Committee to Review Near-Earth Object Surveys and Hazard Mitigation Strategies, Space Studies Board, Aeronautics and Space Engineering Board, Division on Engineering and Physical Sciences, and National Research Council, *Defending Planet Earth: Near-Earth-Object Surveys and Hazard Mitigation Strategies*, 11.

7. A changed ballgame: *Asteroids: Deadly Impact*.

8. Mass hysteria: Ibid.

9. Levy's comment: Ibid.

10. Fragment A's impact: David H. Levy, *Impact Jupiter*, 151.

11. Impact viewing: Ibid., 153–54.

12. Attack on Jupiter: Comet Shoemaker-Levy Collision with Jupiter," NASA, Jet Propulsion Laboratory, last modified March 14, 2000, http://www2.jpl.nasa.gov/sl9/ (accessed March 10, 2014); William J. Board, "When Worlds Collide: A Threat to the Earth Is a Joke No Longer," *New York Times*, August 1, 1994.

13. Moon theories: Ibid., 74–87.

14. Howard Darwin: Dana Mackenzie, *The Big Splat; or, How Our Moon Came to Be*, 2.

15. See's encyclopedia entry: *Encyclopedia Britannica*, 1947 edition (revised 14th ed.), vol. 20 of 24, 273–74.

16. Resisting medium: Ibid., 94.

17. See's blatant meanness: Thomas Jefferson Jackson See, *Researches on the Evolution of the Stellar Systems, Vol. II, The Capture Theory of Cosmical Evolution.*

18. *Nation* book review: Quoted in "Thomas Jefferson Jackson See," *NationMaster.com*, http://www.nationmaster.com/encyclopedia/Thomas-Jefferson-Jackson-See (accessed February 19, 2014).

19. See and Einstein "the fraud": Mackenzie, *Big Splat*, 88–95.

20. The three theories: Ibid., 2.

21. Andesite: Walter Alvarez, *T. rex and the Crater of Doom*, 113.

22. The approach of doom: Ibid., 5.

23. An extreme event: Alvarez, *T. Rex and the Crater of Doom*, 7.

24. Beyond our comprehension: Ibid.

25. Properties of impactors: Ibid., 7–8.

26. Alvarez impact article: Luis W. Alvarez et al., "Extraterrestrial Cause for the Cretaceous-Tertiary Extinction," *Science*, 1095–1108.

27. Arentz: Told to the author.

28. Man on porch: Donald K. Yeomans, *Comets: A Chronological History of Observation, Science, Myth, and Folklore*, 326–27.

29. Burning trees: Ibid.

30. Tunguska explosion: Ibid.

31. The 90 percent: David Morrison, "Impacts and Evolution: Protecting Earth from Asteroids," *Proceedings of the American Philosophical Society*, 439.

32. Morrison on impacts: Ibid.

33. Tyson on asteroid impacts: Neil deGrasse Tyson, *The Sky Is Not the Limit*, 167–68.

34. Survey report and coastal regions: Morrison, "Impacts and Evolution."

35. Planning for NEOs: Clark R. Chapman, Daniel D. Durda, and Robert E. Gold, *The Comet/Asteroid Impact Hazard: A Systems Approach*, 1–2.

36. Alvarez redux: Ibid.

37. American Institute of Aeronautics and Astronautics (AIAA) 2004 Planetary Defense Conference, *White Paper Summarizing Findings and Recommendations from the 2004 Planetary Defense Conference: Protecting Earth from Asteroids.*

38. Spaceguard Foundation established in Rome: "The Spaceguard Foundation Home Page," Spaceguard Foundation, http://spaceguard.rm.iasf.cnr.it/SGF/INDEX.html (accessed February 24, 2014).

39. Spaceguard Foundation's trustees: "Members of the Spaceguard Foundation," Spaceguard Foundation, http://spaceguard.iasf-roma.inaf.it/SGF/members.html (accessed February 24, 2014).

40. Apophis: Ellen Barry, "Russia to Plan Deflection of Asteroid from Earth," *New York Times*, December 31, 2009; Vladimir Isachenkov, "Russia May Send Spacecraft to Knock away Asteroid," Associated Press, December 30, 2009.

41. Possibility of nuclear war: "2002 Eastern Mediterranean Event," *Wikipedia*, last modified February 21, 2014, http://en.wikipedia.org/wiki/Eastern_Mediterranean_event (accessed March 10, 2014).

CHAPTER 3. KNOW THINE ENEMY

1. Sun Tzu on the enemy: Sun Tzu, *The Art of War*, 53.

2. Aristotle's and Pliny's observations: Donald K. Yeomans, *Comets: A Chronological History of Observation, Science, Myth, and Folklore*, 4–5.

3. Pliny and Caesar: Ibid., 13.

4. Sibylline oracles: "NASA: Comets in Ancient Cultures," NASA's John F. Kennedy Space Center, Jet Propulsion Laboratory, and Noah Goldman, University of Maryland, College Park Scholars, http://www.nasa.gov/mission_pages/deepimpact/media/f_ancient.html (accessed February 25, 2014).

5. Han Dynasty: Ibid.

6. Halley and Newton: Mark Littmann and Donald K. Yeomans, *Comet Halley: Once in a Lifetime*, 29.

7. Heaven's Gate: "Heaven's Gate (Religious Group)," *Wikipedia*, last modified February 16, 2014, http://en.wikipedia.org/wiki/ Heaven%27s_Gate_(religious_group) (accessed February 25, 2014).

8. The creation of the Moon: Dana Mackenzie, *The Big Splat; or, How Our Moon Came to Be*, 149–91.

9. Impact craters: Brett Line, "Asteroid Impacts: 10 Biggest Known Hits," *National Geographic News*, February 14, 2013.

10. Explosions and impacts by nation: "List of Meteor Air Bursts," *Wikipedia*, last modified February 28, 2014, http://en.wikipedia.org/ wiki/List_of_meteor_air_bursts (accessed March 11, 2014).

11. Brazilian explosion: John McFarland, "The Day the Earth Trembled," Armagh Observatory, last modified November 10, 2009, http://star.arm.ac.uk/impact-hazard/Brazil.html (accessed February 26, 2014).

12. Sylacauga and Peekskill: "Woman Hit by Falling Star," Xenophilia, August 4, 2001, http://www.xenophilia.com/zb0005 .htm (accessed February 26, 2014); "Peekskill Meteorite," Internet Encyclopedia of Science, Worlds of David Darling, http://www .daviddarling.info/encyclopedia/P/Peekskill_meteorite.html (accessed February 26, 2014).

13. NEAR Shoemaker at 433 Eros: Andrew F. Cheng, "Near Earth Asteroid Rendezvous: Mission Summary," 351.

14. Contemporary asteroid and comet threat: David Morrison, "Impacts and Evolution: Protecting Earth from Asteroids," 439.

15. Pasachoff on impacts: Jay M. Pasachoff, *Contemporary Astronomy*, 395.

16. Keller's thesis: Gerta Keller et al., "Chicxulub Impact Predates the K-T Boundary Mass Extinction," *passim*.

17. Congressional directive to NASA: Spaceguard Survey Report, *Asteroid and Comet Impact Hazards*, sec. 1.2.

18. Congress pushes for international participation: Ibid.

19. Spaceguard and NEAs: Ibid.

20. Worden on asteroids: "Military Perspectives on the Near-Earth Object (NEO) Threat" (speech by Gen. Simon Worden), SpaceRef, US Space Command, July 15, 2002, http://www.spaceref .com/news/viewpr.html?pid=8834 (accessed February 26, 2014).

21. The United Nations conference: John L. Remo, Near-Earth Objects: The United Nations International Conference, *passim*.

22. We're all at equal risk: Spaceguard Survey Report, *Asteroid and Comet Impact Hazards*, sec. 8.1.

23. Congressional testimony: The Threat of Impact by Near-Earth Asteroids, Testimony Before the Subcommittee on Space and Aeronautics of the Committee on Science, May 21, 1998.

24. NASA/Congress/NASA: Eugene Samuel Reich, "NASA Meets Asteroid Discovery Goal," *Nature News Blog*, Nature.com, September 29, 2011, http://blogs.nature.com/news/2011/09/nasa _meets_asteroid_discovery.html (accessed February 26, 2014); Kyle Hill, "NASA Meets Mandate to Identify 90% of Planet-Destroying Asteroids," *Cosmos*, Science News, September 30, 2011.

25. Air Force preface: Maj. Lindley N. Johnson, *Preparing for Planetary Defense: Detection and Interception of Asteroids on Collision Course with Earth*, R-5.

26. Collision prevention: Ibid., 27–30.

27. AIAA position paper: American Institute of Aeronautics and Astronautics (AIAA), "Protecting Earth from Asteroids and Comets," 2–3.

28. An act of Congress: H.R. Rep. No. 108-4544 (2004) (George E. Brown, Jr. Near-Earth Object Survey Act).

29. Johnson's presentation: Lindley Johnson, "Near Earth Object Observations Program Discovers Ten Thousandth NEO."

30. Apophis: Mark Whittington, "NASA Concludes Asteroid Apophis Will Not Hit Earth in 2029 or 2036," Yahoo! News, January 14, 2013, http://news.yahoo.com/nasa-concludes-asteroid-apophis -not-hit-earth-2029-200600191.html (accessed February 26, 2014).

31. Mitigation strategies: *Defending Planet Earth: Near-Earth-Object Surveys and Hazardous Mitigation Strategies*, 66.

32. Arentz: Told to the author.

CHAPTER 4. THE FASCINATION FACTOR

1. Moyer: Michael Moyer, "Eternal Fascinations with the End: Why We're Suckers for Stories of Our Own Demise."

2. Staying the course: See, for example, "Book V," *Bartleby*, http://www.bartleby.com/22/5.html (accessed March 10, 2014).

3. Clarke wrote in a similar vein: Arthur C. Clarke, *Rendezvous with Rama*.

4. Yamamoto review: Judith Yamamoto, review of *Lucifer's Hammer*, by Larry Niven and Jerry Pournelle, *Library Journal*.

5. Amazon review: Alan R. Holyoak, January 19, 2001, customer-review comment on *Lucifer's Hammer*, Amazon.com, http://www.amazon.com/review/RK4E558CDXT5E/ref=cm_cr_dp_title?ie=UTF8&ASIN=0449208133&nodeID=283155&store=books (accessed February 27, 2014). Quotations around film titles have been removed and the titles made italic.

6. *Cleveland Plain-Dealer* review: Editorial reviews for *Lucifer's Hammer* by Larry Niven and Jerry Pournelle, Amazon.com, http://www.amazon.com/Lucifers-Hammer-Larry-Niven-ebook/dp/product-description/B004478DOU/ref=dp_proddesc_0?ie=UTF8&n=133140011&s=digital-text (accessed February 27, 2014). This editorial review appeared on an inside flap of the book.

7. Fictitious impact in Italy: Arthur C. Clarke, *Rendezvous with Rama*.

8. Clarke on Alvarez blurb and Chicxulub: Arthur C. Clarke, *The Hammer of God*, 216–17.

9. The Reborn: Ibid., 165–66.

10. Bo Giertz: Bo Giertz, *The Hammer of God, passim*.

11. Fact with fiction: Clarke, *Rendezvous with Rama*.

12. Resembling a DNA molecule: Clarke, *Hammer of God*, sources and acknowledgments.

13. Kali's attack: Ibid., 160.

14. Kali: Ibid., 211.

15. Clarke's self-deprecation: Ibid., 14.

16. Wolfe's appraisal: Gary K. Wolfe, "The Grand Tours of Arthur C. Clarke," *New York Times* Book Review, March 9, 1997. Quotations around the film title have been removed and the title made italic.

17. *Library Journal* review: Editorial reviews for *The Hammer of God*, by Arthur C. Clarke, Amazon.com http://www.amazon

.com/The-Hammer-God-Arthur-Clarke/dp/product-description/0553
56871X/ref=dp_proddesc_0?ie=UTF8&n=283155&s=books (accessed February 28, 2014). This editorial review refers to the hardcover edition.

18. Not enough science: M. Kidger "bristolcity," June 30, 1998, customer-review comment on *The Hammer of God*, Amazon.com, http://www.amazon.com/review/R2PPAUXINQABB5/ref=cm_cr_pr
_viewpnt#R2PPAUXINQABB5 (accessed February 28, 2014). Quotations around novel titles have been removed and the titles made italic.

19. "I loved this book": "A Customer," March 15, 1997, customer-review comment on *The Hammer of God*, Amazon.com, http://www.amazon.com/review/R1YCHMQT0RT0L7/ref=cm_cr_pr
_viewpnt#R1YCHMQT0RT0L7 (accessed February 28, 2014). Quotations around novel titles have been removed and the titles made italic.

20. Clarke on asteroids: Arthur C. Clarke, *The Exploration of Space*, 135.

21. Tyson on Meteor Crater: Neil deGrasse Tyson, *The Sky Is Not the Limit*, 168–69.

22. Barringer Crater: "Barringer Meteor Crater and Its Environmental Effects," http://www.lpi.usra.edu/science/kring/epo_web/impact_cratering/enviropages/Barringer/barringerstartpage.html (accessed March 3, 2014); "The Barringer Meteorite Crater: A Crater is Born," Barringer Crater Co., http://www.barringercrater.com/about/history_1.php (accessed March 3, 2014).

23. Seventy-nine-page booklet: Dean Smith, *The Meteor Crater Story*.

24. Leonard: Ibid., 63.

25. The lottery analogy: Ibid., 65.

26. *Deep Impact* review: Janet Maslin, "How Do You Reroute a Comet? Carefully," *New York Times*, May 8, 1998. Quotations around film titles have been removed and the titles made italic.

27. Scathing *Armageddon* review: Janet Maslin, "Henny Penny Gets the President's Ear," *New York Times*, July 1, 1998. Quotations around film titles have been removed and the titles made italic.

28. Favorable *Armageddon* review: Michael O'Sullivan, "'Arma-

geddon's' Big Bang Theory," *Washington Post*, July 3, 1998. Quotations around film titles have been removed and the titles made italic.

29. Bruce Willis saves the world: Todd McCarthy, "Review: 'Armageddon,'" *Variety*, June 24, 1998. Quotations around film titles have been removed and the titles made italic.

30. Former teacher review: Jeanine Basinger, "Armageddon," Criterion Collection, June 21, 1999.

31. NASA's disclaimer: *Armageddon*, directed by Jerry Bruckheimer (Burbank, CA: Touchstone Pictures, 1998), 151 min.

32. The number 168: "Feedback: Armageddon Games," *New Scientist*, no. 2619, September 1, 2007.

33. Unrealistic Willis: Gregory Brown, Ben Hall, Ashley Back, and Stuart Turner, "P1_1 Could Bruce Willis Save the World?"

34. Impactor game: "Simulate the Damage Caused by Comet and Asteroid Collisions with *Impact: Earth!*" Open Culture, February 18, 2013, http://www.openculture.com/2013/02/impact_earth.html (accessed March 3, 2014).

35. Nikulin: Andrew E. Kramer, "Step Right up, Kids, the Tiger Will Look Good in Your Photo," *New York Times*, July 11, 2013.

CHAPTER 5. THE OTHER
SALVATION ARMY

1. DSP-647: Robert Lindsey, *The Falcon and the Snowman*, 61–62.

2. Whitman on NEOs: "Year of Meteors. (1859–1860.)," Walt Whitman Archive, http://www.whitmanarchive.org/published/LG/1867/poems/187, 51a.

3. Prophet Miller: Donald K. Yeomans, *Comets: A Chronological History of Observation, Science, Myth, and Folklore*, 178.

4. Carril: Luis Fernández Carril, "The Evolution of Near Earth Objects Risk Perception," *Space Review*, May 14, 2012, http://www.thespacereview.com/article/2080/1 (accessed March 4, 2014).

5. Johnson on planetary defense: Maj. Lindley N. Johnson, *Preparing for Planetary Defense: Detection and Interception of Asteroids on Collision Course with Earth*, R-3–4.

6. Within humanity's grasp: Ibid.

7. Johnson's mitigation strategies: Ibid., R-29.

8. Government involvement: Lt. Col. Rosario Nici and 1st Lt. Douglas Kaupa, "Planetary Defense: Department of Defense Cost for the Detection, Exploration and Rendezvous Mission of Near-Earth Objects," *passim*.

9. A familiar refrain: Ibid.

10. The 1995 UN conference: John L. Remo, Near-Earth Objects: The United Nations International Conference, *passim*.

11. Action Team 14 recommendations: United Nations Office for Outer Space Affairs, "Recommendations of the Action Team on Near-Earth Objects for an International Response to the Near-Earth Object Impact Threat," press handout, February 20, 2013; "Threat of Space Objects Demands International Coordination, UN Team Says," United Nations News Centre, February 20, 2013.

12. Nineteenth-century meteor reports: Andy Newman, "Fireballs in the Sky Are Not Exclusive to Siberia," *New York Times*, February 23, 2013.

13. *Times* Chelyabinsk story: Ellen Barry and Andrew E. Kramer, "Meteor Explodes, Injuring over 1,000 in Siberia," *New York Times*, February 16, 2013.

14. *Times* article, continued: Ibid.

15. Responsible news media: A. C. Charania and Agnieszka Lukaszczyk, *Assessment of Recent NEO Response Strategies for the United Nations*, 4–5.

16. The B612 Foundation and Sentinel: Russell L. Schweickart, "A Call to (Considered) Action"; "Testimony by Ed Lu: Assessing the Risks, Impacts, and Solutions for Space Threats," SpaceRef, March 20, 2013, http://spaceref.com/news/viewsr.html?pid=43620 (accessed March 4, 2014).

17. Lu's testimony: "Testimony by Ed Lu."

18. Sentinel's specs: Robert F. Arentz et al., *NEO Survey: A Fast and Efficient Means for Finding Near-Earth Objects*, *passim*; Sentinel's cost: Bruce Lieberman, "Asteroid Watch," 62.

19. Chapman's testimony: The Threat of Impacts by Near-Earth Asteroids, Testimony before the Subcommittee on Space and

Aeronautics of the Committee on Science, May 21, 1998, 2–4. Quotations around film titles have been removed and the titles made italic.

20. LINEAR: "LINEAR," MIT Lincoln Laboratory, http://www .ll.mit.edu/mission/space/linear/ (accessed March 10, 2014).

21. Planetary Society awards: Planetary Society, "The Planetary Society Takes Central Role in Asteroid Detection, Tracking, and Characterization," press release, April 18, 2013, http://www .planetary.org/press-room/releases/2013/shoemaker-neo-announcement -2013.html (accessed March 4, 2014).

22. Planetary Society probability table and position: Planetary Society, "Sizing Up the Threat from Near-Earth Objects (NEOs)," http://www.planetary.org/explore/space-topics/asteroids-and-comets/ sizing-up-the-threat.html (accessed March 4, 2014).

23. Challenges: "Who We Are," Secure World Foundation, last modified March 4, 2011, http://www.swfound.org/about-us/who -we-are/ (accessed March 4, 2014).

24. Secure World Foundation: "Planetary Defense," Secure World Foundation, last modified July 16, 2013, http://swfound.org/ our-focus/planetary-defense/ (accessed March 4, 2014).

25. Benediktov: Kirill Benediktov, "The Asteroid-Comet Danger and Planetary Defense: A View from Russia."

26. Russia and the United States save the planet: Ibid.

27. The author was on the Survey and Detection Panel.

28. Mitigation techniques: Committee to Review Near-Earth Object Surveys and Hazard Mitigation Strategies, Space Studies Board, Aeronautics and Space Engineering Board, Division on Engineering and Physical Sciences, and National Research Council, *Defending Planet Earth: Near-Earth-Object Surveys and Hazard Mitigation Strategies*, 69–79.

29. A global threat: Ibid.

30. Standing committee and international participation: Ibid., 95–96.

CHAPTER 6. THE DEPARTMENT OF PLANETARY DEFENSE

1. Von Braun and Disney: Mike Wright, "The Disney–von Braun Collaboration and Its Influence on Space Exploration," 2.

2. von Braun's prediction: Cornelius Ryan, *Across the Space Frontier*, 12.

3. Ley on the station: Ibid., 98–117.

4. Haber on survival: Ibid., 82, 97.

5. Bennett's reaction: "Satellite Belittled: Admiral Says Almost Anybody Could Launch 'Hunk of Iron,'" *New York Times*, October 5, 1957.

6. *Times* on Gagarin: "Russians Orbited the Earth Once," *New York Times*, April 13, 1961.

7. Bark sandals: John Noble Wilford, "First into Space, Then the Race," in *Anatomy of the Soviet Union: The Fifty Years*, by Harrison E. Salisbury, 343.

8. Murray: Told to the author.

9. Sorensen on Kennedy and the Moon: Theodore C. Sorensen, *Kennedy*, 525, 105.

10. Speech: John F. Kennedy, "Address at Rice University on the Nation's Space Effort" (Rice University, Houston, Texas, September 12, 1962).

11. Logsdon on Kennedy: John M. Logsdon, *The Decision to Go to the Moon: Project Apollo and the National Interest*, 27–28, 106.

12. The Moon landing story: John Noble Wilford, "Men Walk on Moon," *New York Times*, July 21, 1969.

13. Toynbee and Van Doren: William E. Burrows, "Eminent Thinkers Mull Import of Moon Voyage for Mankind's Future," *Wall Street Journal*, July 18, 1969.

14. Lovell and Mead: Ibid.

15. Asimov: Ibid.

16. The *Times* eats crow: "A Correction," editorial page, *New York Times*, July 17, 1969.

17. Bean: John Noble Wilford, "Giant Leap to Moon, Then Space Lost Allure," *New York Times*, February 9, 2003.

18. Wilson's theory: E. O. Wilson, acceptance speech for the Kistler Prize, quoted in *The Next Thousand Years*, television project by Foundation for the Future, 96.

19. Newitz: Annalee Newitz, "Escape Plans," *Slate*, May 15, 2013, http://www.slate.com/articles/health_and_science/science/2013/05/ surviving_the_next_mass_extinction_humans_will_need_to_leave _earth_for_space.html (accessed March 6, 2014).

CHAPTER 7. THE ULTIMATE STRATEGIC DEFENSE INITIATIVE

1. Chapman on Teller, Wood, and the report: Clark R. Chapman, Daniel D. Durda, and Robert E. Gold, *The Comet/Asteroid Impact Hazard: A Systems Approach*, 12.

2. Association of Space Explorers to the United Nations: Russell L. Schweickart, Thomas D. Jones, Frans von der Dunk, and Sergio Camacho-Lara, *Asteroid Threats: A Call for Global Response*, 3–5.

3. Camacho's statement: "Threat of Space Objects Demands International Coordination, UN Team Says," United Nations News Centre, February 20, 2013.

4. Alvarez article: Luis W. Alvarez et al., "Extraterrestrial Cause for the Cretaceous-Tertiary Extinction," *Science*, 1095–1108.

5. What is required: Spaceguard Survey Report, *Asteroid and Comet Impact Hazards*.

6. LONEOS: Edward Bowell, Bruce Koehn, and Brian Skiff, "The Lowell Observatory Near-Earth-Object Search (LONEOS): Ten Years of Asteroid and Comet Discovery," 4.

7. Spacewatch: NASA, "Near-Earth Object Program: Spacewatch," last modified March 7, 2014, http://neo.jpl.nasa.gov/ programs/spacewatch.html (accessed March 7, 2014).

8. The US reconnaissance satellite program during the Cold War was first code-named Corona and then Keyhole, with successive satellites being called KH-4, KH-5, and so on through KH-11, which sent down imagery in near real time. See William E. Burrows, *Deep Black: Space Espionage and National Security*.

9. Sentinel's cost: Doug Messier, "B612 Foundation's Sentinel Telescope Will Cost $450 Million," *Parabolic Arc*, April 20, 2013, http://www.parabolicarc.com/2013/04/20/b612-foundations-sentinel-telescope-will-cost-450-million/ (accessed March 7, 2014).

10. Sentinel and Lu's statement: Keith Cowing, "B612 Foundation Announces SENTINEL Mission," NASA Watch, June 28, 2012.

11. Musk and Hyperloop: "No Loopy Idea," *Economist*, August 17, 2013, 65.

12. Tonry: University of Hawaii Institute for Astronomy, "ATLAS: The Asteroid Terrestrial-Impact Last Alert System," press release, February 15, 2013, http://www.ifa.hawaii.edu/info/press-releases/ATLAS/ (accessed March 7, 2014).

13. Mainzer: Darren Quick, "NASA's NEOWISE Survey Provides Best Estimate Yet of Potentially Hazardous Asteroids," *Gizmag.com*, May 16, 2012, http://www.gizmag.com/nasa-neowise-pha/22575/ (accessed March 7, 2014).

14. Johnson: Author interview on August 8, 2013.

15. Asteroids have a good side: Ibid.

16. Perozzi: "Target Asteroid Tracked by European Teams," Spaceguard Centre, July 19, 2013, http://www.spaceguarduk.com/news/583-target-asteroid-tracked-by-european-teams (accessed March 7, 2014).

17. Drolshagen: Ibid.

18. Bisei: S. Okumura et al., "Spaceguard Activity in Japan: Past and Future in Bisei Spaceguard Center," *Asteroids, Comets, Meteors* (May 2012), http://www.lpi.usra.edu/meetings/acm2012/pdf/6274.pdf (accessed March 7, 2014).

19. Hayabusa mission: "Hayabusa Spacecraft Returns Asteroid Artifacts from Space," NASA, November 17, 2010, http://www.nasa.gov/topics/solarsystem/features/hayabusa.html (accessed March 7, 2014); "Hayabusa," *Wikipedia*, last modified January 16, 2014, http://en.wikipedia.org/wiki/Hayabusa (accessed March 7, 2014).

20. Chang'e 2: "Chang'e 2," *Wikipedia*, last modified February 23, 2014, http://en.wikipedia.org/wiki/Chang%27e_2 (accessed March 7, 2014).

21. 4179 Toutatis: "4179 Toutatis," *Wikipedia*, last modi-

fied February 20, 2014, http://en.wikipedia.org/wiki/4179_Toutatis (accessed March 7, 2014).

22. "Iron Fist": Jane Perlez, "Chinese, with Revamped Force, Make Presence Known in East China Sea," *New York Times*, July 28, 2013.

23. Puchkov: "FEMA, Russian Ministry to Join Forces against Space Threat," Spaceguard Centre, June 27, 2013, http://www .spaceguarduk.com/news/576-fema-russian-ministry-to-join-forces -against-space-threat (accessed March 7, 2014).

24. Molchanov's statement: "Russia to Use Ballistic Missiles to Fight off Asteroid Threat," *SpaceDaily.com*, January 31, 2013, http://www.spacedaily.com/reports/Russia_to_use_ballistic_missiles _to_fight_off_asteroid_threat_999.html (accessed March 7, 2014).

25. Citadel: "Russia to Build Anti-meteorite Shield—The Project, Titled 'Citadel,' Would Cost about $500 Million, and Could Be Implemented Only with International Cooperation," *InvestmentWatch* (blog), March 12, 2013, http://investmentwatch blog.com/russia-to-build-anti-meteorite-shield-the-project-titled -citadel-would-cost-about-500-million-and-could-be-implemented -only-with-international-cooperation/ (accessed March 7, 2014).

26. Popovkin's reprimand: Doug Messier, "Roscosmos Head Reprimanded for Failures," *Parabolic Arc*, August 2, 2013, http:// www.parabolicarc.com/2013/08/02/49548/ (accessed March 7, 2014).

27. Apollo asteroids are a class of asteroid whose trajectories cross Earth's path. The first one to carry that name was seen and recorded in 1862. They are named after the Greek Sun god because of their close approach to the Sun. The Chelyabinsk asteroid was an Apollo. Suffice it to say they have no connection with the US manned lunar-landing program.

28. AIAA position paper: American Institute of Aeronautics and Astronautics (AIAA), "Dealing with the Threat of an Asteroid Striking the Earth," 6–7.

29. 2004 Planetary Defense Conference position paper: American Institute of Aeronautics and Astronautics (AIAA) 2004 Planetary Defense Conference, *White Paper Summarizing Findings*

and Recommendations from the 2004 Planetary Defense Conference: Protecting Earth from Asteroids, 8–9.

30. 2007 Planetary Defense Conference position paper: American Institute of Aeronautics and Astronautics (AIAA) 2007 Planetary Defense Conference, *White Paper: Summary and Recommendations* 2, 7, 14.

31. AIAA position paper: Space Systems Technical Committee and Systems Engineering Technical Committee, "Responding to the Potential Threat of a Near-Earth-Object Impact," *passim*.

32. Bong Wei's plan: "Asteroid Impact Avoidance," *Wikipedia*, http://en.wikipedia.org/wiki/Asteroid_impact_avoidance (accessed March 7, 2014).

33. IAA position: International Academy of Astronautics (IAA), *2013 IAA Planetary Defense Conference White Paper*, 11–12.

34. Republicans and an asteroid capture: Kenneth Chang, "Plan to Capture an Asteroid Runs into Politics," *New York Times*, July 30, 2013.

35. NASA task force recommendations: NASA Advisory Council, *Final Report of the Ad-Hoc Task Force on Planetary Defense*, October 6–7, 2010, *passim*.

36. Ibid.

37. Morrison: David Morrison, "Impacts and Evolution: Protecting Earth from Asteroids," 444–45.

38. Harris and NEOShield defense: Leonard David, "Asteroid Threat to Earth Sparks Global 'NEOShield' Project," *Space.com*, January 26, 2012, http://www.space.com/14370-asteroid-shield -earth-threat-protection-meeting.html (accessed March 7, 2014).

39. Proponents of SDI maintained that it was stabilizing because it was defensive. Its opponents, the author being among the first of them, believed that it was destabilizing because a nation that had an effective antiballistic-missile capability could be tempted to launch a preemptive first strike against its Cold War adversary because it thought it would survive a counterattack. See William E. Burrows, "Ballistic Missile Defense: The Illusion of Security," *Foreign Affairs*, Spring 1984.

40. DE-STAR: Elizabeth Howell, "Laser-Blasting System Could

Vaporize Big Asteroids," Universe Today, February 15, 2013, http://www.universetoday.com/100021/laser-blasting-system-could-blow-up-big-asteroids/ (accessed March 7, 2014).

41. The need for a survey: David Morrison, Clark R. Chapman, and Paul Slovic, "The Impact Hazard," 87.

42. The Science Definition Team mandate: Near-Earth Object Science Definition Team, Study to Determine the Feasibility of Extending the Search for Near-Earth Objects to Smaller Limiting Diameters: Report of the Near-Earth Object Science Definition Team, iv–v.

CHAPTER 8. THE SURVIVAL IMPERATIVE

1. Schweickart's and Cernan's observations: Frank White, *The Overview Effect: Space Exploration and Human Evolution*, 38–39.

2. Osepok: Buzz Aldrin and John Barnes, *Encounter with Tiber*, 189.

3. The Intruder and the Bombardments: Ibid., 288.

4. The skipper: Told to the author.

5. Barbicane's pronouncement: Jules Verne, *Around the Moon*, 789.

6. The fate of Earth: Ibid., 791.

7. Clarke on human destiny: Arthur C. Clarke, *The Exploration of Space*, 185–86.

8. On making the Moon habitable: Ibid., 111–19.

9. The underground enclave: Ibid., 116.

10. The Moon's advantages: Ibid.

11. Jones on exploration: Tom Jones, *Sky Walking: An Astronaut's Memoir*, 343.

12. Young on NASA: John W. Young, *Forever Young: A Life of Adventure in Air and Space*, 368.

13. Clarke on planetary engineering: Clarke, *Exploration of Space*, 118.

14. Space exploration: Gerard K. O'Neill, "The Colonization of Space."

15. Space colony: He wrote that the cylinders would be about sixteen miles long in one place and twenty miles long in another.

16. O'Neill on colonies: O'Neill, "Colonization of Space," *passim.*

17. Dyson on $96 billion: Freeman Dyson, *Disturbing the Universe*, 124.

18. On responsibility for Island One: Ibid., 124–25.

19. New cultures: William Sims Bainbridge, *Goals in Space: American Values and the Future* of Technology.

20. Bainbridge on colony conditions: Ibid., 167.

21. Spacesuits: Ibid., 112.

22. Murray's vision: Bruce C. Murray, *Journey into Space: The First Thirty Years of Space Exploration.*

23. Murray on financial return: Ibid., 308.

24. Schmitt on the Moon's resources: Harrison H. Schmitt, *Return to the Moon: Exploration, Enterprise, and Energy in the Human Settlement of Space*, 5.

25. On the uses of helium-3: Ibid., 165.

26. Wasser on space development almost stopping: Alan Wasser, "The Space Settlement Initiative," in *Return to the Moon*, edited by Rick N. Tumlinson and Erin R. Medlicott, 105.

27. Wasser and the Space Settlement Initiative: Alan Wasser, "The Space Settlement Initiative," National Space Society, last modified February 2012, http://spacesettlement.org/ (accessed March 8, 2014).

28. Wessen on aviation and spaceflight: Annalee Newitz, Scatter, Adapt, and Remember: How Humans Will Survive a Mass Extinction, 242–43.

29. Newitz on leaving Earth: Ibid., 235.

SELECTED BIBLIOGRAPHY

BOOKS

Aldrin, Buzz, and John Barnes. *Encounter with Tiber*. New York: Warner Books, 1996.

Alvarez, Walter. *T. Rex and the Crater of Doom*. Princeton, NJ: Princeton University Press, 1997.

Bainbridge, William Sims. *Goals in Space: American Values and the Future of Technology*. Albany: State University of New York Press, 1991.

Balmer, Edwin, and Philip Wylie. *When Worlds Collide*. New York: J. B. Lippincott, 1932.

Burrows, William E. *Deep Black: Space Espionage and National Security* (New York: Random House, 1986).

Clarke, Arthur C. *The Exploration of Space*. New York: Harper, 1951.

———. *The Hammer of God*. New York: Bantam Books, 1993.

———. *Rendezvous with Rama*. New York: Bantam Books, 1990.

Dyson, Freeman. *Disturbing the Universe*. New York: Harper and Row, 1979.

Giertz, Bo. *The Hammer of God*. Minneapolis: Augsburg Fortress Publishers, 1960.

Hoyle, Fred, and N. C. Wickramasinghe. *Diseases from Space*. New York: Harper and Row, 1979.

———. *Lifecloud: The Origin of Life in the Universe*. New York: Harper and Row, 1978.

Jones, Tom. *Sky Walking: An Astronaut's Memoir*. New York: Smithsonian-Collins, 2006.

Levine, Arnold S. *Managing NASA in the Apollo Era*. Washington, DC: Scientific and Technical Information Branch, NASA, 1982.

Levy, David H. *Impact Jupiter: The Crash of Comet Shoemaker-Levy 9*. Cambridge, MA: Basic Books, 1995.

Lindsey, Robert. *The Falcon and the Snowman*. New York: Pocket Books, 1979.

Littmann, Mark, and Donald K. Yeomans. *Comet Halley: Once in a Lifetime*. Washington, DC: American Chemical Society, 1985.

Logsdon, John M. *The Decision to Go to the Moon: Project Apollo and the National Interest*. Chicago: University of Chicago Press, 1970.

Mackenzie, Dana. *The Big Splat; or, How Our Moon Came to Be*. New York: John Wiley and Sons, 2003.

McDevitt, Jack. *Moonfall*. New York: HarperPrism, 1998.

McGovern, James. *Crossbow and Overcast*. New York: William Morrow, 1964.

Murray, Bruce C. *Journey into Space: The First Thirty Years of Space Exploration*. New York: W. W. Norton, 1989.

Newitz, Annalee. *Scatter, Adapt, and Remember: How Humans Will Survive a Mass Extinction*. New York: Doubleday, 2013.

Pasachoff, Jay M. *Contemporary Astronomy*. 2nd edition. Philadelphia: CBS College Publishing, 1981.

Remo, John L., ed. *Near-Earth Objects: The United Nations International Conference*. Annals of the New York Academy of Sciences. Vol. 822. New York: New York Academy of Sciences, 1997.

Ryan, Cornelius, ed. *Across the Space Frontier*. New York: Viking, 1952.

Saint-Exupéry, Antoine de. *The Little Prince*. New York: Harvest, 1943.

Salisbury, Harrison E., ed. *Anatomy of the Soviet Union*. London: Nelson, 1967.

Schmitt, Harrison H. *Return to the Moon: Exploration, Enterprise, and Energy in the Human Settlement of Space*. New York: Copernicus Books, 2006.

See, Thomas Jefferson Jackson. *Researches on the Evolution of the Stellar Systems, Vol. II, The Capture Theory of Cosmical Evolution*. Lynn, MA: T. P. Nichols, 1896.

Shapiro, Robert. *Origins: A Skeptic's Guide to the Creation of Life on Earth*. New York: Bantam, 1987.

Smith, Dean. *The Meteor Crater Story*. Flagstaff, AZ: Meteor Crater Enterprises, 1996.

Sorensen, Theodore C. *Kennedy*. New York: Harper and Row, 1965.

Steel, Duncan. *Rogue Asteroids and Doomsday Comets: The Search for the Million Megaton Menace That Threatens Life on Earth*. New York: John Wiley and Sons, 1995.

Sun Tzu. *The Art of War*. Translated by Thomas Cleary. Boston: Shambhala, 1988.

Tumlinson, Rick N., with Erin R. Medlicott, eds. *Return to the Moon*. Burlington, ON: Apogee Books, 2005.

Tyson, Neil deGrasse. *The Sky Is Not the Limit: Adventures of an Urban Astrophysicist*. Amherst, NY: Prometheus Books, 2004.

Verne, Jules. *Around the Moon*. In *The Jules Verne Omnibus* (four volumes in one), by Jules Verne. Philadelphia: J. B. Lippincott, 1973.

Wells, H. G. *The War of the Worlds*. New York: Pocket Books, 1988.

White, Frank. *The Overview Effect: Space Exploration and Human Evolution*. Boston: Houghton Mifflin, 1987.

Yeomans, Donald K. *Comets: A Chronological History of Observation, Science, Myth, and Folklore*. New York: John Wiley and Sons, 1991.

———. *Near-Earth Objects: Finding Them before They Find Us*. Princeton, NJ: Princeton University Press, 2013.

Young, John W., with James R. Hansen. *Forever Young: A Life of Adventure in Air and Space*. Gainesville: University Press of Florida, 2012.

BOOK CHAPTERS

Morrison, David, Clark R. Chapman, and Paul Slovic. "The Impact Hazard." In *Hazards Due to Comets and Asteroids*, by Tom Gehrels. Tucson: University of Arizona Press, 1994.

ARTICLES

Alvarez, Luis W., Walter Alvarez, Frank Asaro, and Helen V. Michel. "Extraterrestrial Cause for the Cretaceous-Tertiary Extinction." *Science* 208, no. 4448 (June 6, 1980): 1095–1108.

Benediktov, Kirill. "The Asteroid-Comet Danger and Planetary Defense: A View from Russia." *Executive Intelligence Review*, April 26, 2013.

Bowell, Edward, Bruce Koehn, and Brian Skiff. "The Lowell Observatory Near-Earth-Object Search (LONEOS): Ten Years of Asteroid and Comet Discovery." *Astronomical Review* (May 21, 2011).

Brown, Gregory, Ben Hall, Ashley Back, and Stuart Turner. "PI_1 Could Bruce Willis Save the World?" *Journal of Physics Special Topics* 10, no. 1 (November 1, 2011).

Cheng, Andrew F. "Near Earth Asteroid Rendezvous: Mission Summary." *Johns Hopkins APL Technical Digest* 19, no. 2 (1998).

Johnson, Lindley. "NASA's Near-Earth Object Observations Program Discovers Ten Thousandth NEO." SciTechDaily.com, June 26, 2013. http://scitechdaily.com/nasas-near-earth-object-observations -program-discovers-ten-thousandth-neo/ (accessed February 21, 2014).

Keller, Gerta, Thierry Adattee, Wolfgang Stinnesbeck, Mario Rebolledo-Vieyra, Jaime Urrutia Fucugauchi, Utz Kramar, and Doris Stüben. "Chicxulub Impact Predates the K-T Boundary Mass Extinction." *Proceedings of the National Academy of Sciences* 101, no. 11 (March 16, 2004).

Lieberman, Bruce. "Asteroid Watch." *Air and Space Magazine*, January 2013.

Morrison, David. "Impacts and Evolution: Protecting Earth from Asteroids." *Proceedings of the American Philosophical Society* 154, no. 4 (December 2010).

Moyer, Michael. "Eternal Fascinations with the End: Why We're Suckers for Stories of Our Own Demise." In "The End." Special issue, *Scientific American* (September 2010).

Nici, Lt. Col. Rosario, and lst Lt. Douglas Kaupa. "Planetary Defense:

Department of Defense Cost for the Detection, Exploration, and Rendezvous Mission of Near-Earth Objects." *Airpower Journal* (summer 1997).

O'Neill, Gerard K. "The Colonization of Space." *Physics Today*, September 1974.

Von Braun, Wernher. "Exploration to the Farthest Planets." *New Scientist* 22, no. 387, April 16, 1964.

Wright, Mike. "The Disney–von Braun Collaboration and Its Influence on Space Exploration." In *Selected Papers from the 1993 Southern Humanities Conference*. Edited by D. Schenker, C. Hanks, and S. Kray. Huntsville, AL: Southern Humanities Press, 1993.

Yamamoto, Judith T. Review of *Lucifer's Hammer*, by Larry Niven and Jerry Pournelle. *Library Journal* 102, no. 13 (July 1, 1977).

REPORTS

American Institute of Aeronautics and Astronautics (AIAA) 2004 Planetary Defense Conference. *White Paper Summarizing Findings and Recommendations from the 2004 Planetary Defense Conference: Protecting Earth from Asteroids*. Washington, DC: AAAS, February 2004.

American Institute of Aeronautics and Astronautics (AIAA) 2007 Planetary Defense Conference. *White Paper: Summary and Recommendations*. Washington, DC: NASA, May 10, 2007.

Arentz, Robert F., et al. *NEO Survey: A Fast and Efficient Means for Finding Near Earth Objects*. Ball Aerospace and Technologies Corp., September 8, 2010.

Chapman, Clark R., Daniel D. Durda, and Robert E. Gold. *The Comet/Asteroid Impact Hazard: A Systems Approach*. Boulder, CO: Southwest Research Institute, February 24, 2001.

Charania, A. C., and Agnieszka Lukaszczyk. *Assessment of Recent NEO Response Strategies for the United Nations*. Spaceworks Engineering, 2009.

Committee to Review Near-Earth Object Surveys and Hazard Mitigation Strategies, Space Studies Board, Aeronautics and Space

Engineering Board, Division on Engineering and Physical Sciences, and National Research Council. *Defending Planet Earth: Near-Earth-Object Surveys and Hazard Mitigation Strategies.* Washington, DC: National Academies Press, 2010.

International Academy of Astronautics (IAA). *2013 IAA Planetary Defense Conference White Paper.* Flagstaff, AZ: IAA, April 2013.

Johnson, Maj. Lindley N. *Preparing for Planetary Defense: Detection and Interception of Asteroids on Collision Course with Earth.* White paper in *Spacecast 2020.* Washington, DC: Air University, US Air Force, 1993.

NASA Advisory Council, *Final Report of the Ad-Hoc Task Force on Planetary Defense.* October 6–7, 2010.

Near-Earth Object Science Definition Team. *Study to Determine the Feasibility of Extending the Search for Near-Earth Objects to Smaller Limiting Diameters: Report of the Near-Earth Object Science Definition Team.* Washington, DC: NASA, August 22, 2003.

McAndrew, James. *The Roswell Report: Case Closed.* Washington, DC: US Air Force, US Government Printing Office, 1997.

Schweickart, Russell L., Thomas D. Jones, Frans von der Dunk, and Sergio Camacho-Lara. *Asteroid Threats: A Call for Global Response.* Edited by Jessica Tok. Association of Space Explorers International Panel on Asteroid Threat Mitigation, September 25, 2008.

Spaceguard Survey Report. *Asteroid and Comet Impact Hazards.* Washington, DC: NASA, 1992.

SCHOLARLY AND OTHER PAPERS AND PRESENTATIONS

American Institute of Aeronautics and Astronautics (AIAA). "Dealing with the Threat of an Asteroid Striking the Earth." Position paper, American Institute of Aeronautics and Astronautics, Reston, VA, April 1990.

American Institute of Aeronautics and Astronautics (AIAA). "Protecting Earth from Asteroids and Comets, American Institute of

Aeronautics and Astronautics: An AIAA Position Paper." Position paper, American Institute of Aeronautics and Astronautics, Reston, VA, October 2004.

Launius, Roger D. "Why Go to the Moon? The Many Faces of Lunar Policy." Article presented at the American Astronautical Society meeting, Greenbelt, MD, March 17, 2004.

Schweickart, Russell L. "A Call to (Considered) Action." Presentation at the National Space Society International Space Development Conference, Washington, DC, May 20, 2005.

Space Systems Technical Committee and Systems Engineering Technical Committee. "Responding to the Potential Threat of a Near-Earth-Object Impact: An AIAA Position Paper." Position paper, American Institute of Aeronautics and Astronautics, Reston, VA, September 1995.

Tagliaferri, Edward. "The History of AIAA's Interest in Planetary Defense." AIAA paper 96-4381, AIAA Meeting Papers on Disc, September 1996.

The Threat of Impact by Near-Earth Asteroids, Testimony Before the Subcommittee on Space and Aeronautics of the Committee on Science. 105th Cong. (May 21, 1998). Statement of Clark R. Chapman, Southwest Research Institute.

FILM

Asteroids: Deadly Impact. Directed by Eitan Weinreich. Washington, DC: National Geographic Video, 2003. DVD, 60 min.

INDEX

ABOUT THE AUTHOR

William E. Burrows, professor emeritus of journalism and mass communication at New York University, is the author of twelve books and numerous articles in the *New York Times*, the *Washington Post*, the *Wall Street Journal*, *Foreign Affairs*, and other publications. His previous books include *The Survival Imperative: Using Space to Protect Earth*; *By Any Means Necessary: America's Secret Air War in the Cold War*; *The Infinite Journey: Eyewitness Accounts of NASA and the Age of Space*; *This New Ocean: The Story of the First Space Age*; *Mission to Deep Space: Voyagers' Journey of Discovery*; *Exploring Space: Voyages in the Solar System and Beyond*; and *Deep Black: Space Espionage and National Security*. He is the coauthor (with Robert Windrem) of *Critical Mass: The Dangerous Race for Superweapons in a Fragmenting World*. Burrows was the only nonscientist on the National Research Council's Near-Earth Object Survey and Detection Panel. In recognition of his distinguished career and expertise, NASA has named a Main Belt asteroid after him, and he is a recipient of the American Astronautical Society John F. Kennedy Astronautics Award, among other honors.